"十三五"高等职业教育规划教材

智能可穿戴设备的设计与实现

魏 彦　孙宏伟　主　编
杨 欧　连国云　副主编
李金子　史 虹　参　编

中国铁道出版社有限公司
CHINA RAILWAY PUBLISHING HOUSE CO., LTD.

内容简介

本书以智能可穿戴设备开发平台为依托，并结合编者的实际开发经验编写而成，内容包括：可穿戴设备的现状、平台简介和实验说明；IAR-EWARM 软件的安装和使用方法，用 Android Studio 开发软件开发可穿戴设备 APP；Android 入门基础知识；编程实现让 APP 通过蓝牙和设备进行通信；可穿戴设备模块综合设计。

本书结构清晰，讲解细致，注重学生实践能力的培养。通过学习，读者可对智能可穿戴设备的设计与实现有总体认识，并掌握基本的开发方法。

本书适合作为高等职业院校计算机专业的教材，也可作为对可穿戴设备硬件开发感兴趣的相关人士的参考用书。

图书在版编目（CIP）数据

智能可穿戴设备的设计与实现/魏彦，孙宏伟主编 .—北京：中国铁道出版社有限公司，2019.8（2023.7重印）

"十三五"高等职业教育规划教材

ISBN 978-7-113-25838-2

Ⅰ.①智… Ⅱ.①魏… ②孙… Ⅲ.①移动终端-智能终端-设计-高等职业教育-教材 Ⅳ.①TN87

中国版本图书馆 CIP 数据核字（2019）第 115158 号

书　名：智能可穿戴设备的设计与实现
作　者：魏　彦　孙宏伟

策　划：王春霞　尹　鹏　　　　　　　编辑部电话：（010）63551006
责任编辑：王春霞　彭立辉
封面设计：付　巍
封面制作：刘　颖
责任校对：张玉华
责任印制：樊启鹏

出版发行：中国铁道出版社有限公司（100054，北京市西城区右安门西街 8 号）
网　　址：http://www.tdpress.com/51eds/
印　　刷：三河市航远印刷有限公司
版　　次：2019 年 8 月第 1 版　2023 年 7 月第 2 次印刷
开　　本：850 mm×1 168 mm　1/16　印张：13.25　字数：277 千
书　　号：ISBN 978-7-113-25838-2
定　　价：36.00 元

版权所有　侵权必究

凡购买铁道版图书，如有印制质量问题，请与本社教材图书营销部联系调换。电话：（010）63550836
打击盗版举报电话：（010）63549461

前言

随着医疗和消费领域对智能可穿戴设备的需求与日俱增,智能可穿戴设备的应用得到了极大的普及。目前,智能手表、智能手环的应用十分广泛;未来,在健康监护、家庭康复、安全监测、疾病早发现、疗效评估等领域,智能可穿戴设备必将得到更加广泛的应用。

智能穿戴设备所涉及的技术包括人工智能技术、核心传感技术、通信技术、设备交互技术等,是一种综合性很强的设备。对于广大智能可穿戴设备的开发人员而言,要求的知识领域广,入门较难,专业书籍较少。为此,编者结合实际开发经验,精选开发平台,以实际案例为基础,编写了本书,旨在为广大有志于从事智能可穿戴设备开发的人员提供入门的便捷方式。

从知识结构上看,智能可穿戴设备的开发,主要涉及的知识包括传感器与单片机技术、操作系统编程技术。本书以一套智能可穿戴设备开发平台为依托,以可穿戴设备中常用模块为案例,从传感器技术和 Android 智能可穿戴 APP 程序设计两方面,为读者详尽讲解了智能可穿戴设备的开发方法。

本书主要涉及以下几方面内容:

1. Android 程序设计基础知识

本书案例中,编译器使用 IAR;Android 编程使用 Android Studio。本书对这两个工具的安装和使用进行了简单介绍,并针对 Android 零基础的学习者,提供了 Android 入门基础知识。对于已有 Android 基础的开发者,可以跳过这部内容。

2. 可穿戴设备开发通用技术

可穿戴设备开发通用技术主要包括网络通信技术、蓝牙通信技术、Fragment 方法、Activity 间通信方法、通信协议的设计方法等。这些技术在可穿戴设备的开发中应用非常广泛,是每一个可穿戴设备开发者和学习者必须要掌握的通用技术。该部分内容有详尽的代码讲解,对于初学者而言,能够从详尽的代码讲解中获取到通用的设计方法。

3. 常用模块系统设计

常用智能可穿戴设备模块综合设计,主要包括体温采集模块、脚底压力模块、心率计数模块、计步模块、腕部触感提醒模块、紫外线超测量模块等常用模块。每个模块从传感器的选择开始,进行电路原理的设计,再到 Android 程序的编写,介绍了每个模块的设计流程。

在学完成这些典型模块后,有志于智能可穿戴设备开发的广大开发人员和学生,可以根据这个设计流程,制定设计思路,选择传感器模块,设计硬件电路图,编写单片机程序和 Android 程序,进行其他功能模块的设计开发。

目前,智能可穿戴设备应用开发方面的教材非常少,本书的创新之处在于结合教师教学和科研实践,从通用技术到具体模块,从入门到综合设计,深入浅出地介绍了智能可穿戴设备开发的流程和知识体系。

1. 精心规划内容

本书的编者长期从事嵌入式方向的教学,对可穿戴设备的开发亦有丰富经验。在编写过程中,精心规划了教学内容,并与一线开发工程师研讨,内容满足可穿戴设备开发初学者必须掌握的技能。

2. 精讲通用技术

通用技术为各种类型的可穿戴设备开发中较为普遍、应用较多的技术,本书将其作为一部分精讲内容,让读者能够在充分掌握后,应用到今后可能从事的各种各样的智能可穿戴设备的开发中。

3. 精选典型案例

本书选择的案例为智能可穿戴设备中使用频率较高的典型案例,可指导读者开发可穿戴设备。

本书是在深圳职业技术学院计算机应用专业开设两届的"智能可穿戴设备应用开发"课程的基础上,根据实际课程教学的需求,经过计算机应用专业相关教师严谨的调研、充分的实践、不断的讨论,精心编撰而成。深圳职业技术学院计算机应用专业在嵌入式方向的教学中,不断改革创新,课程设置因时而变,依托深圳改革创新的大环境,为培养嵌入式领域的优秀人才而努力。在此特别感谢计算机应用专业全体教师为本书的撰写付出的艰苦努力。

本书提供课程教学大纲、教学进度计划表、全书案例代码。如果需要相关教材资料,或者对教材有宝贵意见和建议,请发送邮件至作者邮箱 37475569@qq.com。

由于时间仓促,编者水平有限,书中难免存在疏漏和不妥之处,恳请广大读者朋友不吝赐教。

编 者

2019 年 3 月

目 录

第1章 绪论 1

1.1 可穿戴设备的现状及应用 1
　1.1.1 可穿戴设备定义 1
　1.1.2 可穿戴设备起源 2
　1.1.3 可穿戴设备典型应用分析 2
　1.1.4 可穿戴设备崛起 3
1.2 可穿戴设备应用开发系统介绍 3
1.3 可穿戴实验平台使用说明 4
　1.3.1 平台介绍 4
　1.3.2 Android APP 6
小结 10
习题 10

第2章 IAR-EWARM 软件开发入门 11

2.1 IAR-EWARM 安装步骤 11
　2.1.1 在线注册 11
　2.1.2 下载软件 13
　2.1.3 安装 EWARM 14
2.2 新建软件工程 15
　2.2.1 创建工程 16
　2.2.2 添加文件 17
　2.2.3 配置工程 19
2.3 基于 STMEVKIT-STM32F10××8 的示例代码运行 23

小结 24
习题 24

第3章 Android Studio 基础 25

3.1 Android Studio 的环境搭建 25
　3.1.1 Windows 7 64 位系统安装 JDK 及环境变量配置 25
　3.1.2 Android Studio 安装及工程创建 30
　3.1.3 Android Studio 的简单设置 34
3.2 创建第一个 Android 程序 43
　3.2.1 创建第一个 Android 项目 43
　3.2.2 熟悉 Android 项目的目录结构 46
　3.2.3 在平板电脑中运行 Android APP 47
　3.2.4 Android 常用组件 50
　3.2.5 Android 常用布局 52
3.3 Android 用户登录及注册实验 53
　3.3.1 登录界面 UI 设计 53
　3.3.2 注册界面 UI 设计 56
　3.3.3 注解控件 ButterKnife 的使用 59

I

3.3.4 实现 Android 与服务器间的 HTTP 通信 ………… 59
3.4 Android 程序签名打包 ………… 64
3.5 Android TextView(文本框)详解 ………… 67
3.6 EditText(输入框)详解 ………… 72
 3.6.1 设置默认提示文本 ………… 72
 3.6.2 获得焦点后全选组件内所有文本内容 ………… 73
 3.6.3 限制 EditText 输入类型 ………… 73
 3.6.4 设置最小行、最多行、单行、多行、自动换行 ………… 74
 3.6.5 设置文字间隔及英文字母大写类型 ………… 75
 3.6.6 控制 EditText 四周的间隔距离与内部文字与边框间的距离 ………… 75
 3.6.7 设置 EditText 获得焦点，同时弹出小键盘 ………… 75
 3.6.8 EditText 光标位置的控制 ………… 76
 3.6.9 带表情的 EditText 的简单实现 ………… 77
 3.6.10 带删除按钮的 EditText ………… 78
3.7 ImageView(图像视图) ………… 80
 3.7.1 src 属性和 background 属性的区别 ………… 81
 3.7.2 解决 blackground 拉伸导致图片变形的方法 ………… 82
 3.7.3 src 和 background 结合应用 ………… 83
 3.7.4 adjustViewBounds 设置缩放是否保存原图长宽比 ………… 84

3.7.5 scaleType 设置缩放类型 ………… 85
小结 ………… 88
习题 ………… 88

第 4 章 可穿戴实验平台蓝牙通信设计 ………… 89

4.1 Android 菜单界面创建与 Fragment 使用 ………… 89
 4.1.1 Fragment 的创建 ………… 89
 4.1.2 GridView 的使用 ………… 91
 4.1.3 Fragment 的切换 ………… 95
4.2 Android 蓝牙 4.0 通信与通信协议设计 ………… 96
 4.2.1 Android 蓝牙 4.0 详解 ………… 96
 4.2.2 蓝牙 4.0 开发 ………… 96
 4.2.3 可穿戴蓝牙的数据通信 ………… 102
4.3 Android 蓝牙开发——搜索蓝牙设备 ………… 106
 4.3.1 打开蓝牙设备 ………… 107
 4.3.2 搜索蓝牙设备 ………… 108
 4.3.3 获取配对过的蓝牙设备 ………… 109
4.4 Android 蓝牙开发——连接蓝牙设备 ………… 109
小结 ………… 112
习题 ………… 112

第 5 章 可穿戴设备模块综合设计 ………… 113

5.1 体温检测模块的开发和设计 ………… 114
 5.1.1 体温采集信息价值 ………… 114
 5.1.2 非接触式体温传感器原理 ………… 115
 5.1.3 温度采集电路解析 ………… 116
 5.1.4 体温传感代码解析 ………… 116

5.1.5 手机 APP 软件的开发和功能 …………… 118
5.2 可穿戴脚底压力模块 …………… 129
  5.2.1 电阻应变式传感器原理及应用 …………… 129
  5.2.2 拉力传感器与压力传感器 …………… 130
  5.2.3 力敏传感器应用电路 …………… 131
  5.2.4 拉力与压力程序解析 …………… 132
  5.2.5 手机 APP 软件的开发和功能 …………… 134
5.3 心率计数模块 …………… 145
  5.3.1 心率采集信息价值 …………… 145
  5.3.2 非接触式体温传感器原理 …………… 146
  5.3.3 心率传感器电路解析 …………… 147
  5.3.4 心率传感代码解析 …………… 147
  5.3.5 手机 APP 软件的开发和功能 …………… 151
5.4 手腕佩戴式计步模块 …………… 160
  5.4.1 运动传感器原理及发展历程 …………… 160
  5.4.2 MPU 6050 传感器的使用 …………… 161
  5.4.3 步伐识别算法 …………… 162
  5.4.4 运动传感代码解析 …………… 164
  5.4.5 手机 APP 软件的开发和功能 …………… 166
5.5 腕部触感提醒项目 …………… 173
  5.5.1 微型振动马达原理及使用注意事项 …………… 173
  5.5.2 微型振动马达电路解析 …………… 174
  5.5.3 振动提示程序解析 …………… 175
  5.5.4 手机 APP 软件的开发和功能 …………… 178
5.6 紫外线超测量模块 …………… 185
  5.6.1 紫外线的检测原理及注意事项 …………… 185
  5.6.2 紫外线传感器电路解析 …………… 187
  5.6.3 紫外线传感代码解析 …………… 188
  5.6.4 手机 APP 软件的开发和功能 …………… 192
小结 …………… 203
习题 …………… 203

第1章 绪 论

本章主要介绍可穿戴设备的现状、应用、起源等理论知识,然后介绍所用到的设备开发系统,最后介绍所用到的实验平台。

学习目标

- 了解可穿戴设备的现状及应用。
- 熟悉可穿戴设备应用开发系统。
- 学会使用可穿戴应用开发系统。

1.1 可穿戴设备的现状及应用

1.1.1 可穿戴设备定义

可穿戴设备也称可穿戴计算设备,目前并没有统一的定义。麻省理工学院的媒体实验室对可穿戴计算的定义是:"计算机科技结合多媒体和无线传播以不突显异物感的输入或输出仪器(如首饰、眼镜或衣服)进行连接个人局域网络功能、侦测特定情境或成为私人智慧助理,进而成为使用者在行进动作中处理信息的工具"。常见的可穿戴设备的几种应用类型,如图1-1所示。

基于这个定义,可穿戴设备可理解为基于人体自然能力之上的,借助计算机(习称电脑)科技实现对应业务功能的设备。人体自然能力指人类本体与生俱来的能力,如动

图1-1 可穿戴设备的几种应用类型

手能力、行走能力、语言能力、眼睛转动能力、心脏脉搏跳动能力、大脑神经思维能力等；这里的电脑科技指基于人体能力或环境能力通过内置传感器、集成芯片功能实现对应的信息智能交互功能。

1.1.2 可穿戴设备起源

最早的可穿戴设备可追溯到便携式计算器时代，如以腕表的方式集成计算器的功能方便人们随时随地进行简单的数字运算。随着技术的进步，越来越多的计算功能被集成到可穿戴设备上。从穿戴式电脑到穿戴式设备，业界对可穿戴的定义进一步明确，认为可穿戴设备需要包含"可穿戴的形态""独立的计算能力""专用的程序或功能"这三类基本属性。例如，2006年Eurotech公司推出的手腕式电阻触屏电脑如图1-2所示，符合传统对可穿戴设备的定义。但这样的设备消费者并不买账，很大一方面就是类似的设备用户体验并不友好，另外也没有凸显和传统电脑的区别。

可见，除了上述三类基本属性外，可穿戴设备最重要的一个属性是"用户体验"友好，可融入人体自然穿戴中，穿戴起来不显唐突。另外，要突出和传统电脑的区别，一个差异属性就是"感知能力"，即通过内置芯片传感器可感知穿戴者或周围环境属性信息，为业务处理提供更多的基础信息输入。随着物联网、传感技术、芯片技术和智能操作系统的发展，可穿戴设备和人体结合程度越来越紧密，用户体验也更加友好，更多的可穿戴设备被投入市场。图1-3所示为目前流行的智能穿戴手环。

图1-2 手腕式电阻触屏电脑

图1-3 智能穿戴手环

1.1.3 可穿戴设备典型应用分析

按照联合国人口老龄化标准，全球现已进入老龄社会，40年后全球老龄化程度与目前发达国家相当。健康和医疗领域被公认为可穿戴设备最有发展潜力、市场规模最大的一个领域。健康领域的可穿戴设备以轻量化的手表、手环和配饰为主要形式，实现运动或户外数据（如心率、步频、卡路里消耗等指标）的检测、分析与服务。医疗领域的可穿戴设备以专业化方案提供血压、心率等医疗体征的检测和处理，形式较为多样，包括医疗手表、手机附件等。

医疗健康类可穿戴设备（如智能手环等）通过内置传感器采集人体指标并通过蓝牙方式把

数据传递给智能手机内对应的 APP 应用;APP 应用通过数据通信网络把数据传递到对应的数据处理中心;数据中心在接收到数据后进行业务逻辑处理,对于一些异常指标数据可以按照业务逻辑触发医务人员进行进一步的分析,处理后反馈给用户。可穿戴设备定期采集人体各项指标数据并进行记录,这个过程中医务人员通过远程查看方式及时掌控各项指标,实现对生命全过程的健康监控和常见疾病的防御。

我国患病人数多、医疗成本高、患病时间长、服务需求大,合理的慢性疾病管理,能够避免看急诊和住院治疗,减少就医次数,可以大大节约费用和人力成本。医疗健康类可穿戴设备从某种程度来说符合很多人的刚性需求,未来将会更多地融入人们的生活。

1.1.4 可穿戴设备崛起

2013 年是可穿戴设备崛起之年,谷歌眼镜、三星智能手表等设备的发布标志着互联网时代硬件创新达到了新的顶峰,这些用户体验更为友好的可穿戴设备使人们的身体也将由此成为一个智能终端,并影响人们的行为模式,提高人们的行动效率,最终将改变人们接入互联网的方式和入口。

在互联网时代,人与计算机设备交互周期通常以天为单位,人们可能每天使用几次计算机进行网络冲浪;移动互联网时代解放了人们的双腿,人们每间隔几个小时就会使用一次手机进行信息查阅,人机交互周期缩短到以小时为单位;而在可穿戴计算时代,人与设备的交互周期将以分钟、秒为单位,交互频率更大。移动互联网用户规模较桌面互联网爆炸式增长事实说明业务的交互越频繁、业务的用户黏性就越大,进而促进用户规模的扩大。基于此可穿戴时代用户规模或比移动互联网更大,这也是资本市场垂青可穿戴设备的原因所在。随着越来越多可穿戴设备的出现,这些设备的形态、应用类型各式各样,需要对其进行归类以更好地了解其特征和技术特点。

1.2 可穿戴设备应用开发系统介绍

可穿戴设备实训平台针对可穿戴设备进行全生态实践教学,所授课内容涉及物联网技术、电子工程、计算机工程、移动互联网、通信工程、生物医疗电子等多项内容,尤其适合物联网技术、电子工程、计算机软件、医疗电子、通信工程等多个专业的课程教学。实训系统如图 1-4 所示。

该实训系统中除了包含腕带、腰带等多种可穿戴设备外,还配备有专门定制的智能手表,如图 1-5 所示。该智能手表可搭载相关应用,实现穿戴设备间的数据采集与数据传输。

该实训平台针对实际应用和相关传感器,可进行多种类型的可穿戴设备实验;既可进行单一传感器实验,也可以多传感器组合进行更为复杂的监测。

智能可穿戴设备的设计与实现

图1-4 实训系统

图1-5 智能手表实物图

1.3 可穿戴实验平台使用说明

1.3.1 平台介绍

可穿戴设备可直接在配套实训系统中使用,当实训系统上电后,可在显示屏上看到可穿戴设备的数据;可通过点击屏幕上的"振动"按钮来控制马达设备振动。

脑电波、压力、拉力设备需要外接检测装置才能获取准确的数据,请在设备上电前外接好检测装置,确保数据的准确性。如果在上电后外接检测装置,则显示的数据有可能不准确。

实训系统配件如图1-6所示,包括可穿戴腰带、可穿戴头带、可穿戴脚踏、可穿戴绑带各1件。

可穿戴头带间距连接线6条,如图1-7所示。

图1-6 实训系统配件

图1-7 可穿戴头带间距连接线

AUX 对录线 2 条,如图 1-8 所示。

图 1-8　AUX 对录线

其他设备:USB 充电线 1 条;ST-LINK/V2　1 个;安卓平板电脑(7 英寸)1 个。

可穿戴设备可通过连接线与蓝牙设备连接,蓝牙设备带独立锂电池,可为可穿戴设备模块供电。使用手机 APP 连接蓝牙设备时,蓝牙设备名称为 UART-××××××,符合上述名称规范的蓝牙设备均可连接。蓝牙设备与可穿戴设备接线方式如图 1-9 所示。

图 1-9　蓝牙设备与可穿戴设备接线方式

蓝牙设备可级联多个可穿戴设备模块,如图 1-10 所示。

图 1-10　级联多个可穿戴设备模块

部分传感器(如压力传感器、脑电波传感器等)可通过 AUX 对录线与可穿戴设备模块进行连接,如图 1-11 所示。

图 1-11　压力传感器、脑电波传感器及 AUX 对录线

1.3.2 Android APP

实训系统中配备 Android 平板电脑,可通过平板电脑中的"可穿戴技术平台"APP 与可穿戴模块进行数据交互。如果 Android 平板电脑没有安装"可穿戴技术平台"APP 可通过 https://www.pgyer.com/WuLI 进行下载。

Android APP 可通过蓝牙设备操作 8 种可穿戴设备,分别是心率、马达、加速度、温度、脑电波、拉力、压力和紫外线设备。可通过点击"连接设备"弹出的选择框,选择需要连接的蓝牙设备。蓝牙设备名称:UART－××××××,符合上述名称规范的蓝牙设备均可连接。图 1 － 12 所示为 APP 连接界面。

1. 心率设备

心率设备可通过与蓝牙设备连接,把检测的心率数据返回给 APP。在 APP 主界面中点击"心率",可跳转到心率数据界面,如图 1 － 13 所示。系统可实时显示采集到的心率数据,也可以通过点击"历史折线"进入历史数据界面,查看连接以来采集到的数据折线。

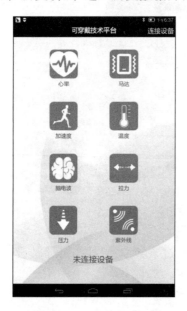

图 1 －12　APP 蓝牙连接界面　　　　图 1 －13　心率数据界面

2. 马达设备

马达设备可通过与蓝牙设备连接,实现振动次数或振动时间的控制。在 APP 主界面中点击"马达"可跳转至马达数据界面,其控制方式可分为"连续振动"和"短促振动"两种,都能控制马达设备的振动,其数据界面如图 1 － 14 所示。

3. 加速度设备

加速度设备可通过与蓝牙设备连接实现对加速度数据的获取。在 APP 主界面中点击"加速度",可跳转至加速度数据界面,如图 1 － 15 所示。如果带着加速度设备行走,则加速度设备记录行走的步数,同时把采集到的步数数据返回给 APP,在 APP 端可以看到步数数据的增加。

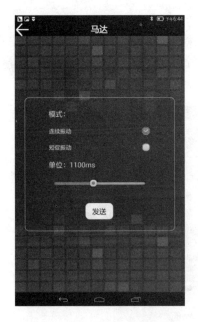 

图 1-14 马达数据界面　　　　　图 1-15 加速度数据界面

4. 温度设备

温度设备可通过与蓝牙设备连接,实现对温度数据的获取。在 APP 主界面中点击"温度",可跳转至温度数据界面。在温度界面中可实时显示采集到的温度数据,也可以通过点击"历史折线"进入历史数据界面,查看连接以来采集到的数据折线。其数据界面如图 1-16 所示。

图 1-16 温度数据界面

5. 脑电波设备

脑电波设备可通过与蓝牙设备连接，实现对脑电波数据的获取。在 APP 主界面中点击"脑电波"，可跳转至脑电波数据界面。在脑电波数据界面中，可实时显示采集到的脑电波数据。其数据界面如图 1-17 所示。

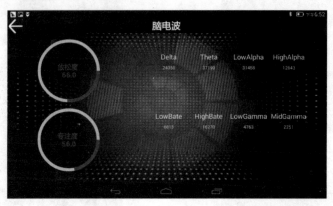

图 1-17　脑电波数据界面

注意：使用脑电波设备前请先连接脑电波设备的外接检测装置，确保数据的准确性。

6. 拉力设备

拉力设备可通过与蓝牙设备连接，实现对拉力数据的获取。在 APP 主界面中点击"拉力"，可跳转至拉力数据界面。在拉力数据界面中，可实时显示采集到拉力设备的数据。也可以通过点击"历史折线"进入历史数据界面，查看连接以来采集到的数据折线数据。其数据界面如图 1-18 所示。

图 1-18　拉力数据界面

注意：使用拉力设备前请先连接拉力设备的外接检测装置，确保数据的准确性。

7. 压力设备

压力设备可通过与蓝牙设备连接，实现对压力数据的获取。在 APP 主界面中点击"压力"，可跳转至压力数据界面。在压力数据界面中，可实时显示采集到压力设备的数据。也可以通过点击"历史折线"进入历史数据界面，查看连接以来采集到的数据折线。其数据界面如图 1-19 所示。

注意：使用压力设备前请先连接压力设备的外接检测装置，确保数据的准确性。

8. 紫外线设备

紫外线设备可通过与蓝牙设备连接，实现对紫外线数据的获取。在 APP 主界面中点击"紫外线"，可跳转至紫外线数据界面。在此紫外线的界面图中，可实时显示采集到紫外线设备的数据。也可以通过点击"历史折线"进入历史数据界面，查看连接以来采集到的数据折线。其数据界面如图 1-20 所示。

图 1-19　压力数据界面

图 1-20　紫外线数据界面

注意：紫外线设备需要在阳光的照射下方可获取紫外线当前的强度值。

模块与模块之间可以通过 IIC 级联进行通信，但级联的头模块必须是蓝牙模块，因为蓝牙模块提供了供电和总线通信数据的处理发送。

传感模块和震动模块不能同时级联，否则通信将出错。

小 结

　　本章对常见的可穿戴设备做了介绍，列举了典型的例子，并且对本书中所需要用到的可穿戴实验平台进行了简单了解。

习 题

　　1. 现实中你见过哪几种可穿戴设备？它们的通信方式分别是什么？
　　2. 本章介绍的可穿戴设备由哪几部分构成？分别有什么作用？

第 2 章

IAR-EWARM 软件开发入门

可穿戴实验平台在底层硬件上采用 ARM 芯片作为主控芯片,配套的开发环境为 IAR Embedded Workbench for ARM,该开发环境是 IAR Systems 公司为 ARM 微处理器开发的一个集成开发环境(以下简称 IAR EWARM)。相比其他的 ARM 开发环境,IAR EWARM 具有入门容易、使用方便和代码紧凑等特点。

IAR EWARM 中包含一个全软件的模拟程序(simulator)。用户不需要任何硬件支持就可以模拟各种 ARM 内核、外围设备甚至中断的软件运行环境,从中可以了解和评估 IAR EWARM 的功能和使用方法。

学习目标

- 能够安装 IAR EWARM 软件。
- 能够正确地使用 IAR EWARM 创建、打开软件工程。
- 能够运行 demo。

2.1 IAR-EWARM 安装步骤

2.1.1 在线注册

在线注册可以通过 http://supp.iar.com/Download/SW/? item = EWARM - KS32,下载 EWARM KickStart 版本(32KB 代码大小限制),此版本免费使用。其界面如图 2 - 1 所示。

单击 Continue 按钮并填写注册信息,单击 Submit Registration 按钮完成本步骤,如图 2 - 2 所示。

图 2-1　下载界面

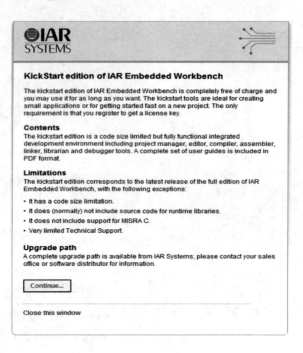

图 2-2　注册界面

2.1.2 下载软件

下载软件的步骤如下：

(1) 注册后，在注册信息中提供的信箱将会自动收到一封邮件，如图 2-3 所示。

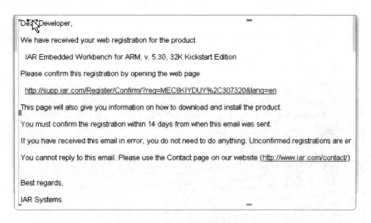

图 2-3 收到的邮件

(2) 单击邮件中的链接，打开注册码界面，如图 2-4 所示。

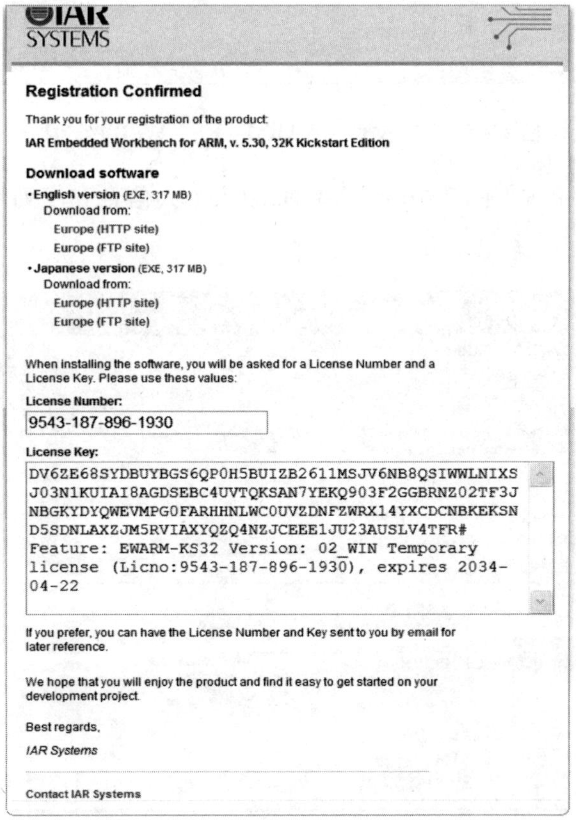

图 2-4 注册码界面

(3)注册后就可以下载安装 EWARM KickStart 程序,可以选择 HTTP 下载,也可以选择 FTP 下载。将安装程序保存到本地磁盘。

2.1.3 安装 EWARM

安装 EWARM 的操作步骤如下:

(1)双击已经下载的安装文件 EWARM – KS – WEB – 5302,选择 Install IAR…选项,如图 2 – 5 所示。

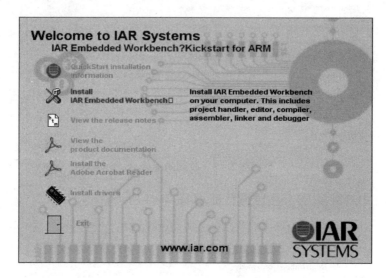

图 2 – 5　安装文件 EWARM – KS – WEB – 5302

(2)在安装过程中,出现如图 2 – 6 所示界面,填入个人信息和序列号。序列号在接收到的邮件中可以获得。

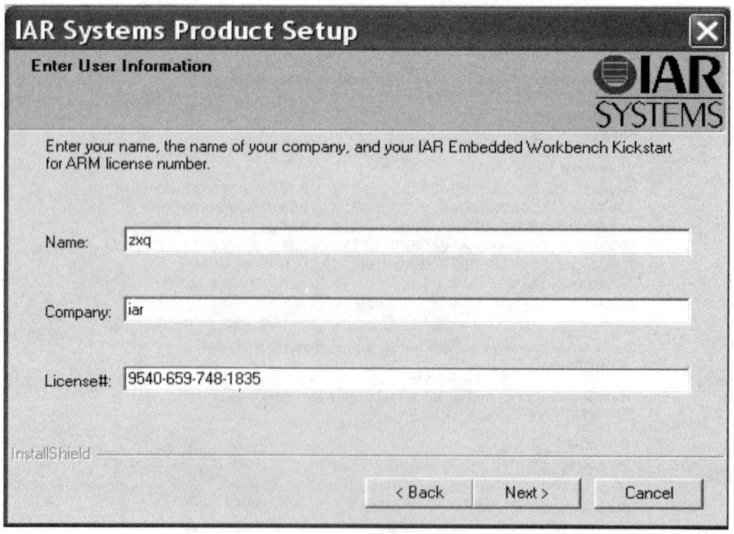

图 2 – 6　填写用户名信息

(3) 单击 Next 按钮,继续安装,出现如图 2-7 所示界面,在接收的邮件中输入序列号。

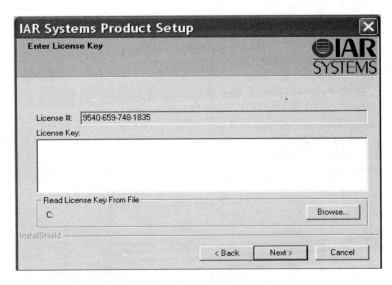

图 2-7　License Key 填写界面

(4) 继续安装过程,直到出现如图 2-8 所示界面,单击 Finish 按钮,结束安装过程。

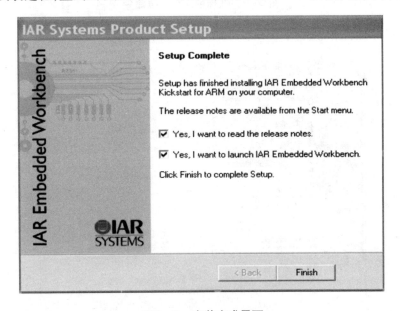

图 2-8　安装完成界面

2.2　新建软件工程

新建一个简单、基础的软件工程需要 3 个步骤:创建工程、添加文件和配置工程。

上面的准备工作做好之后,即可新建自己的软件工程。下面将详细讲述新建一个软件工程的详细过程。

2.2.1 创建工程

创建工程的操作步骤如下:

(1)打开软件,选择 Project→Create New Project 命令,如图2-9所示。

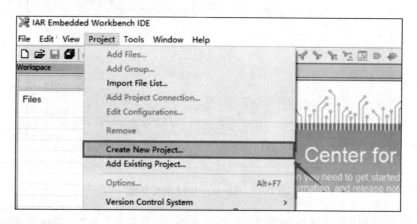

图2-9　选择创建新工程命令

(2)在弹出的对话框中选中 Empty project,单击 OK 按钮创建工程,如图2-10所示。

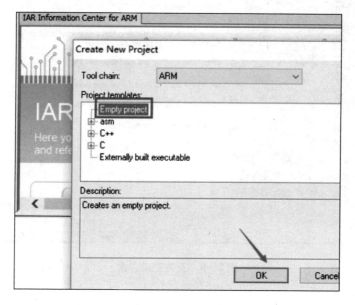

图2-10　创建新工程

(3)创建好工程后,会出现如图2-11所示界面,需要进一步添加文件到工程和配置工程。

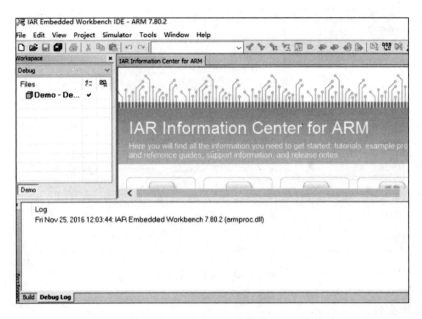

图 2-11　空工程界面

2.2.2　添加文件

添加文件就是将源代码(前面提取的库、新建的文件等)添加到工程中。

1. 工程中添加组

在 IAR 软件中右击工程中的选项,选择 Add→Add Group 命令,如图 2-12 所示。

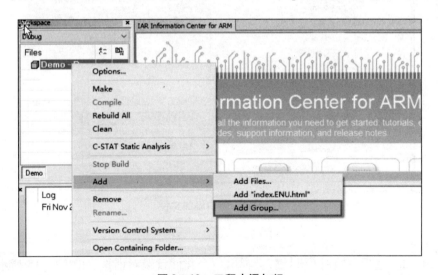

图 2-12　工程中添加组

2. 填写组的名称

在弹出的对话框中填写组的名称,单击 OK 按钮,如图 2-13 所示。

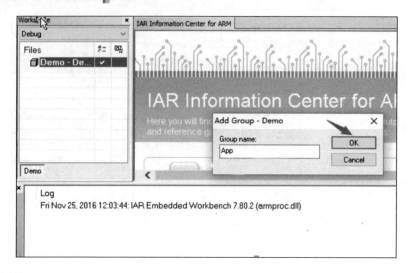

图 2-13　填写组的名称

3. 组中添加文件

在创建完成的组中,在 App 文件夹继续添加文件。右击 App 文件夹,选择 Add→Add Files 命令,如图 2-14 所示。

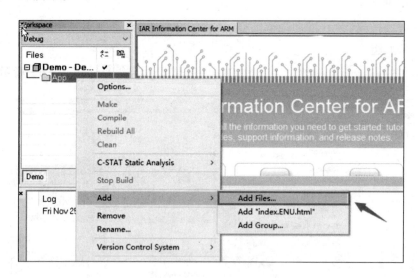

图 2-14　组中添加文件

4. 选择添加文件

选中 Add Files 选项,按住【Ctrl】键,会弹出一个文件目录对话框,如图 2-15 所示。在对话框中选中要添加的文件,单击"打开"按钮。

5. 文件添加完成

把需要添加的文件按照上面的步骤,添加到工程中,可以显示在左侧项目文件列表,直到添加完成,如图 2-16 所示。

第 2 章　IAR-EWARM 软件开发入门

图 2-15　选中要添加的文件

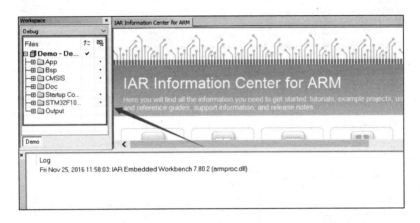

图 2-16　添加完成界面

2.2.3　配置工程

配置工程对于初学者来说，大部分内容只需要默认即可，这里只讲述几个常见的配置，能满足基本的功能。

1. 配置选项

在左侧项目结构目录中右击项目名称，选择 Options 命令设置配置选项，如图 2-17 所示。

图 2-17　选择配置项

2. 选择器件

在弹出的 Options 对话框中选择 Target，选中 Device 单选按钮，然后单击右侧的按钮选择器件名称，如图 2-18 所示。

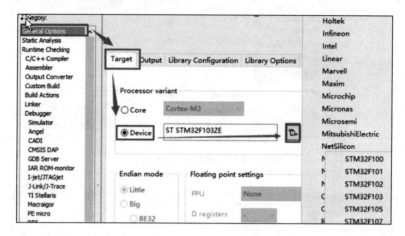

图 2-18 选择器件

3. 库配置

如果需要使用某些标准的库函数接口（如 printf 和 scanf），就需要选择 Library 下的 Full 选项，如图 2-19 所示。

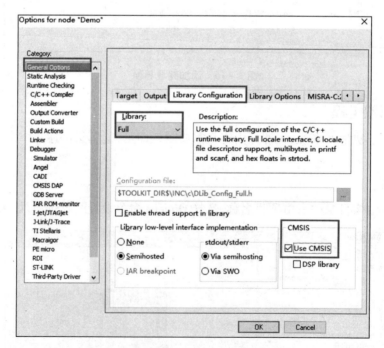

图 2-19 库配置

选中 CMSIS（微控制器软件接口标准）下的 Use CMSIS 复选框。

4. 预处理——添加路径

添加的路径最好是相对路径，而不是绝对路径。使用绝对路径工程位置改变之后就找不到文件，就会出错。可以单击路径选择按钮选择路径，也可以通过复制文件路径进行配置。选择 C/C++ Compiler 选项，再选择属性中的 Preprocessor，最后单击 ··· 按钮（见图 2-20），一步一步添加路径，直到最后完成。

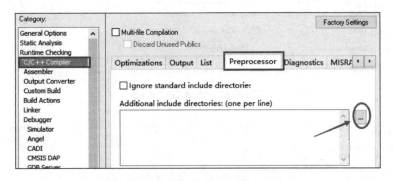

图 2-20 添加路径

5. 预定义

这里的预定义类似于在源代码中的 #define ××× 这种宏定义。图 2-21 中的 STM32F10X_HD 在 stm32f10x.h 中打开即可，选中 STM32F10X_HD。

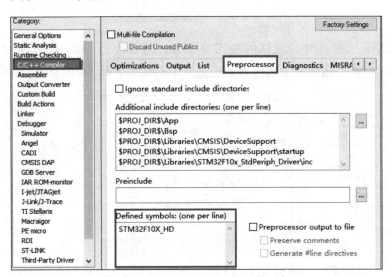

图 2-21 选中 STM32F10X_HD

6. 输出 Hex 文件

选择 Output Converter 选项，再选择 Intel extended 即可，如图 2-22 所示。

7. 选择下载调试工具

选择 Debugger 选项，单击 Setup，再选择 Driver 编号即可，如图 2-23 所示。

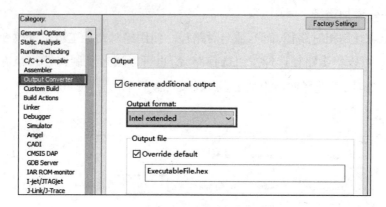

图 2-22 填写 Hex 名称

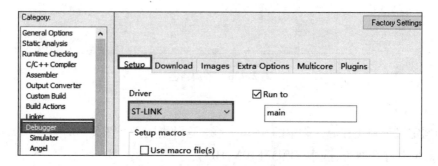

图 2-23 选择调试工具

使用 ST-LINK 工具有时其默认的接口是 JTAG,需要改为 SWD 才能使用。具体操作如下:选择 ST-LINK,在右侧弹出的属性界面中选择 Interface 下的 SWD 即可,如图 2-24 所示。

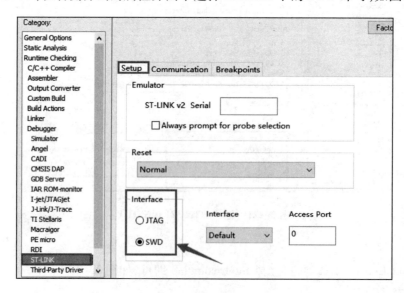

图 2-24 下载调试工具选择 SWD

2.3 基于 STMEVKIT-STM32F10××8 的示例代码运行

1. 硬件配置

(1) STLink_USB:跳线选择 ST-LINK。

(2) USB 线:连接 CN4(ST-LINK)和 PC。

2. 运行 Demo 文件

操作步骤如下:

(1) 打开 EWARM 集成开发环境中的 Workspace:选择 File→Open→Workspace 命令在 Open Workspace 对话框中,在…\IAR-STMEVKIT\GPIO 路径下,选择文件 GPIO Demo.eww,单击 Open 按钮打开 Workspace 文件,如图 2-25 所示。

图 2-25 打开 Workspace 文件

(2) 编译和连接项目:在 Workspace 面板中,右击项目名 GPIO Demo,选择 Rebuild All 命令编译和连接所有的项目文件,如图 2-26 所示。也可以通过选择 Project→Rebuild All 命令来实现相同的功能。完成此步骤后,在 Build 面板中将提示没有警告和错误的信息。

(3) 下载程序:单击工具栏中的 按钮或者选择 Project→Debug 命令下载代码到 Android Studio 空间。调试器将会在 main() 函数的入口处停止。

(4) 调试和执行程序:关于程序调试的详细信息,请参考 EWARM_UserGuide.pdf 的 Part.4 Debugging 章节(也可以选择 Help 菜单,打开 ARM Embedded Workbench User Guide)。这里单击工具栏中的 按钮运行程序。

智能可穿戴设备的设计与实现

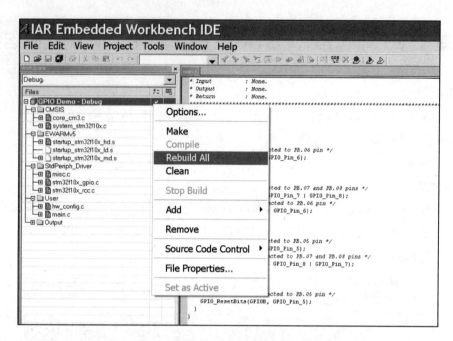

图 2-26 Build 项目

  小　　结

本章介绍了如何下载安装 IAR EWARM 软件,并且使用 IAR EWARM 软件创建和配置工程,如何基于 STMEVKIT - STM32F10××8 运行 Demo。

  习　　题

1. 了解有无和 IAR EWARM 软件功能相似的一些软件。
2. 编写一个程序,通过 STLink_USB 和 EWARM 下载到开发板中,观察运行情况。

第 3 章

Android Studio 基础

本章主要围绕学习 Android Studio 相关的软件开发知识,通过编写一些实例程序来学习 Android 组件。

学习目标

- 能够正确对 Android Studio 进行环境搭建。
- 可以用 Android Studio 创建项目。
- 实现项目的登录和注册功能。
- 掌握 Android 基本组件的开发。

3.1 Android Studio 的环境搭建

学习 Android 开发的第一步是 Java 环境 JDK 的搭建和 Android Studio 软件的安装。Android 程序开发用的是 Java 语言,因此需在电脑上配置 JDK(Java Development Kit)环境。

3.1.1 Windows 7 64 位系统安装 JDK 及环境变量配置

1. 安装 JDK 开发环境

(1)下载网站(http://www.oracle.com/)界面如图 3-1 所示,单击 Downloads 图标进入下载界面。

(2)在弹出的界面中选择 JDK Download 下载 JDK,如图 3-2所示。

(3)在弹出的界面中,选中 Accept License Agreement 单选按钮同意相关许可,选择 Windows x86 版本进行下载,如图 3-3所示。

智能可穿戴设备的设计与实现

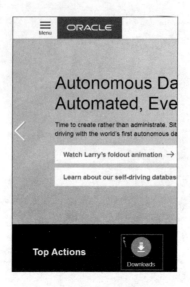

图 3-1　下载网站截图

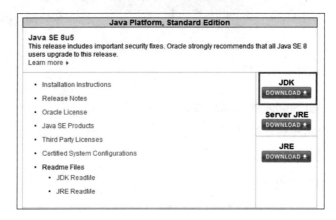

图 3-2　下载 JDK

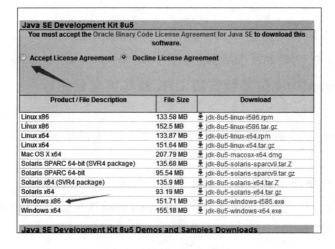

图 3-3　同意相关许可

(4)打开下载完成后的安装包,在弹出界面中,选择 Development Tools 开始安装 JDK,可以单击"更改"按钮选择路径,如图 3-4 所示。

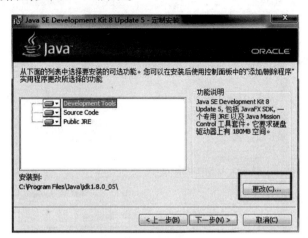

图 3-4　安装 JDK

(5)单击"更改"按钮后,在界面中修改文件夹名,单击"确定"按钮修改即可,如图 3-5 所示。

图 3-5　修改安装目录

(6)确定之后,单击"下一步"按钮,继续安装过程。

注意:当提示安装 JRE 时,可以选择不要安装。

2. 配置环境变量

对于 Java 程序而言,主要会使用 JDK 的两个命令:javac.exe、java.exe,路径为 C:\Java\jdk1.8.0_05\bin。但是,这些命令由于不属于 Windows 自己的命令,所以要想使用,还需要进行路径配置。

右击"计算机"图标,选择"属性"命令,选择"高级系统设置"选项,在弹出的"系统属性"对话框中选择"高级"选项卡,单击"环境变量"按钮,如图3-6所示。

图3-6 创建环境变量

在"环境变量"栏下单击"新建"按钮,创建新的系统环境变量,如图3-7所示。

图3-7 环境变量操作

具体设置方法可参考Java程序设计的书籍,这里不再详细介绍。

3. 确认环境配置是否正确

在控制台分别输入java、javac、java - version命令,会出现JDK的编译器信息,包括修改命令的语法和参数选项等信息。

(1) 输入 java 命令后出现的信息如图 3-8 所示。

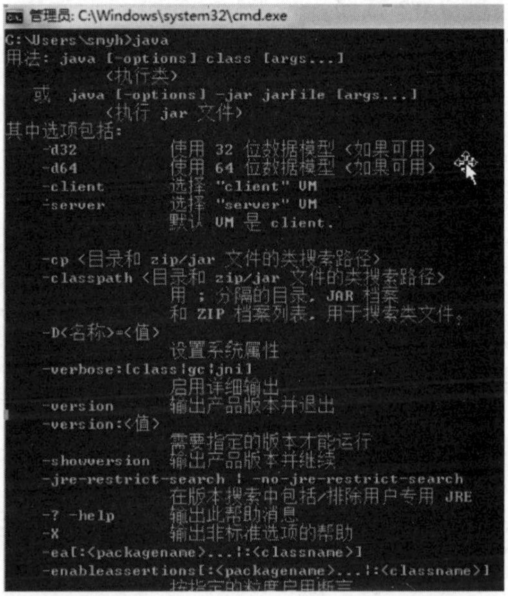

图 3-8 输入 java 命令后出现的信息

(2) 输入 javac 命令后出现的信息如图 3-9 所示。

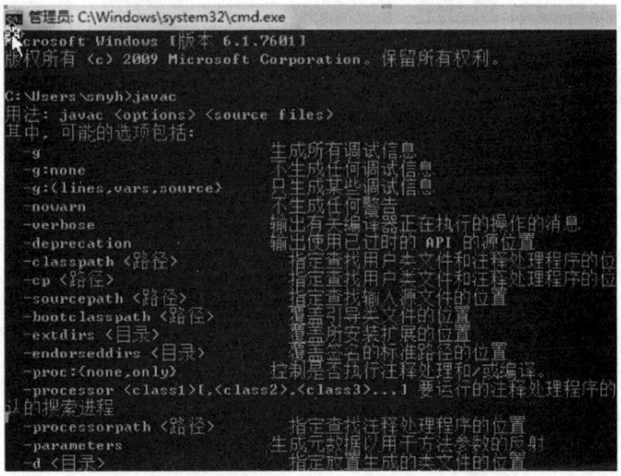

图 3-9 输入 javac 命令后出现的信息

(3) 输入 java - version 命令后出现的信息如图 3-10 所示。

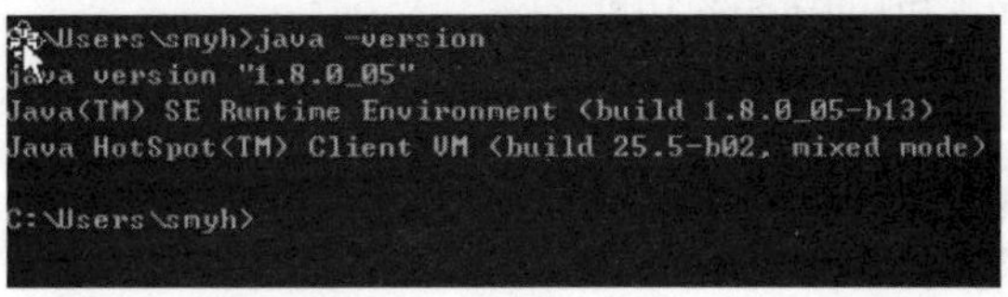

图 3-10 输入 java - version 命令后出现的信息

至此配置步骤已经完成,接下来开始验证 Java 程序。

4. 在控制台下验证第一个 Java 程序

```
publiccl Android Studios Test {
publicstaticvoid main(String[] args) {
   System.out.println("Hello Java");
   }
}
```

用记事本编写程序后,单击"保存"按钮,并存入 C 盘根目录后,输入 javac Test.java 和 java Test 命令,即可运行程序(打印出结果"Hello Java")。

程序解析:

(1)编写 Java 源代码程序,扩展名.java。

(2)在命令行模式中,输入命令:javac 源文件名.java,对源代码进行编译,生成 Android Studios 字节码文件。

(3)编译完成后,如果没有报错信息,输入命令:java Test,对 clAndroid Studios 字节码文件进行解释运行,执行时不需要添加.clAndroid Studios 扩展名,如图 3 – 11 所示。

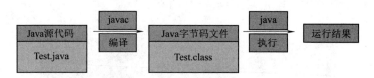

图 3 – 11 程序编译流程图

注意:若输入 javac Test.java 命令后,显示 Javac 不是内部或外部命令,原因是因为没有提前安装好 JDK 开发环境或环境变量配置有误。

3.1.2 Android Studio 安装及工程创建

Android Studio 的优势:

(1)基于 Gradle 的构建支持。

(2)Android 特定重构和快速修复。

(3)更加丰富的模板代码,让创建程序更简单。

(4)提示工具更好地对程序性能、可用性、版本兼容和其他问题进行控制捕捉。

(5)直接支持 ProGuard 和应用签名功能。

(6)自带布局编辑器,可以让用户拖放 UI 组件,并在多个屏幕上预览布局等。

(7)内置 Google 云支持。

(8)内置 svn、git 工具支持。

(9)支持插件,eclipse 有的 Android Studio 中基本都能找到。

(10) Android Studio 2.0 之后,支持 NDK。

1. 第一次安装

Android Studio 安装完成后,第一次启动 Android Studio 前,为了避免重新下载新版本的 SDK,需要做如下操作:

(1) Android Studio 启动前,打开安装目录,请先将 bin 目录的 idea.properties 文件中增加一行:disable.android.first.run = true,避免第一次打开 Android Studio 时自动重新下载 SDK。在 Mac 平台,右击安装包,选择 Show Package Contents,即可找到 bin 目录。

(2) 第一次打开 Android Studio 时,需要配置 JDK 和 SDK,单击 Configure 下拉按钮,在弹出的下拉列表中选择 Project Defaults→Project Structure 命令,如图 3-12 所示。

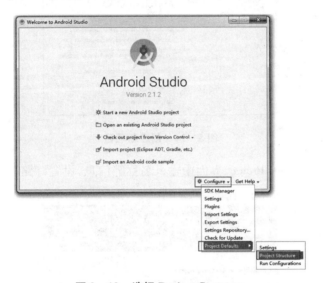

图 3-12　选择 Project Structure

(3) 在弹出的 Project Structure 界面的 SDK Location 文本框中填写具体路径,保存即可,如图 3-13 所示。

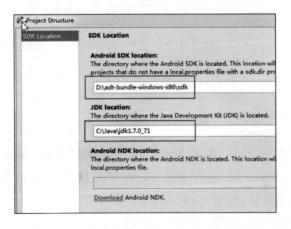

图 3-13　修改 JDK 和 Android SDK 的路径

在使用 Android Studio 时,也可随时修改 JDK 和 Android SDK 的路径。选择 File→Other Settings→Default Sructure,即可进行同样的修改。

2. 新建一个 Android Studio 工程

(1)配置好 JDK 和 Android SDK 后,就可以开始新建 Android 项目。双击 Start a new Android Studio project,如图 3 – 14 所示。

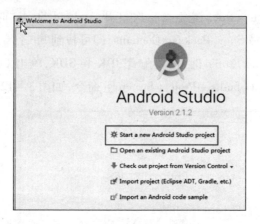

图 3 – 14　创建新项目

(2)在弹出的 Create New Project 界面中输入名称等信息,设置好路径,如图 3 – 15 所示。

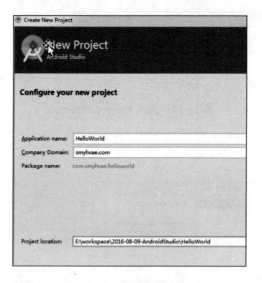

图 3 – 15　设置项目名称

(3)设置好工程名和路径之后,单击 Next 按钮,即可新建完成,如图 3 – 16 所示。

第一次安装 Android Studio 时,会出现提示安装 SDK 的窗口,在此界面中会自动安装 SDK,只需要耐心等待安装完成即可,如图 3 – 17 所示。

(4)下载完成后,单击 Next 按钮,如图 3 – 18 所示。

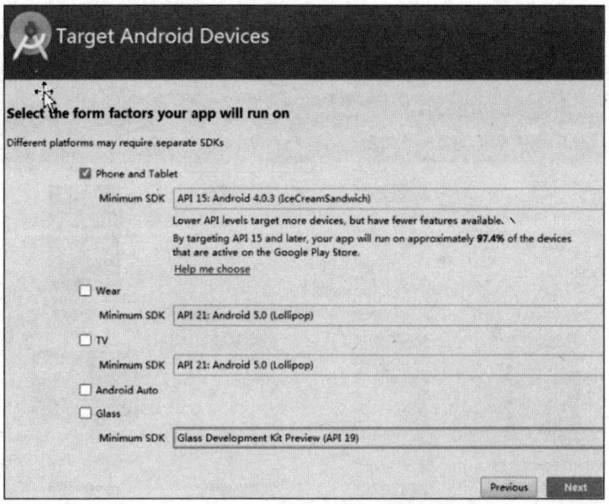

Android Studio 基础　第 3 章

图 3-16　选择 Android 设备

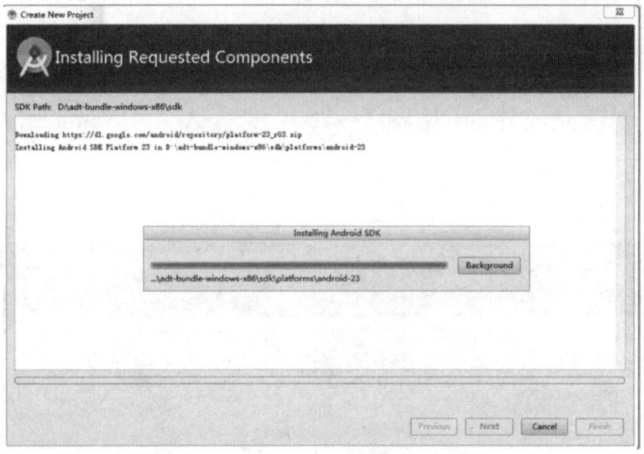

图 3-17　自动安装 SDK

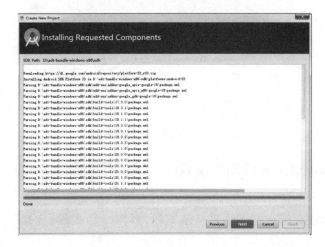

图 3-18　SDK 下载完成

(5) 当出现如图 3-19 所示的界面后,选择创建项目类型,双击 Empty Activity。

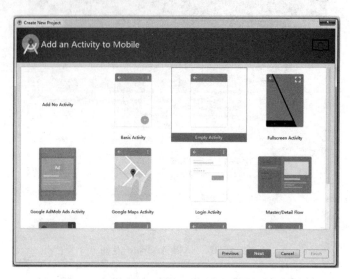

图 3-19　选择创建项目类型

(6) 在弹出的项目创建完成界面(见图 3-20),填写 Actity Name 和 Layout Name,然后单击 Finish 按钮,应用即可创建完成。

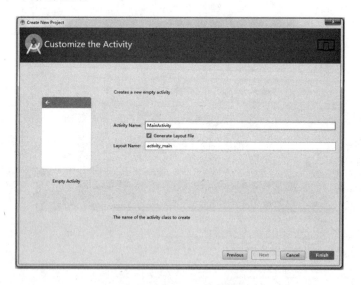

图 3-20　项目创建完成界面

3.1.3　Android Studio 的简单设置

1. 主题修改

Android Studio 软件支持自定义修改风格主题,具体操作步骤:选择 File→Settings 命令,在弹出的对话框中选择 Apperance→Theme→Darcula 选项,然后单击 OK 按钮即可,如图 3-21 所示。

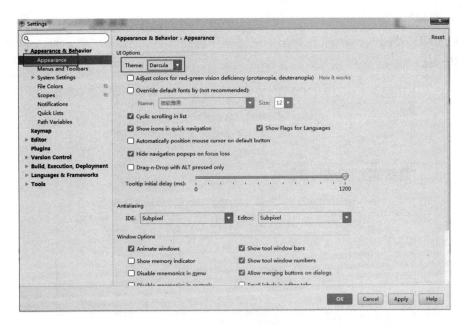

图 3-21　修改主题风格

2. 导入第三方主题

如果不喜欢系统提供的几种主题，可以进入网站 http://color-themes.com/ 来获取第三方主题，直接进行下载后，是一个 jar 包，回到 Android Studio，选择 File→Import Settings 命令，将下载的 jar 包导入即可，如图 3-22 所示。

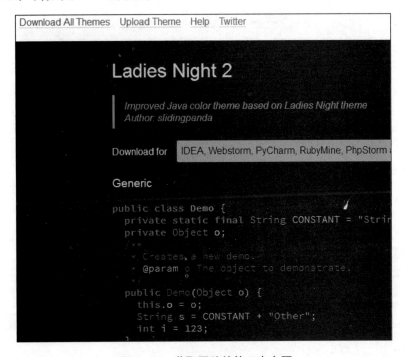

图 3-22　获取网站的第三方主题

3. 代码字体修改

(1)选择 File→Settings 命令,在弹出的对话框中选择 Editor→Colors & Fonts→Font 选项,修改控制台的字体,如图 3-23 所示。

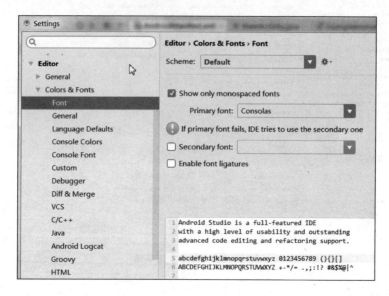

图 3-23　修改控制台的字体

(2)修改完之后发现 Android Studio 的一些默认字体(如侧边栏的工程目录的字体)并没有发生变化。修改 Android Studio 的默认字体,可选择 Console Font→Scheme,如图 3-24 所示。

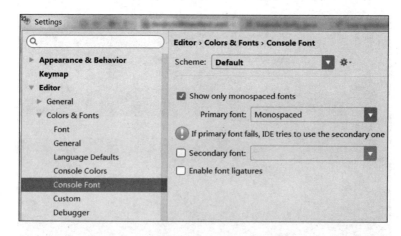

图 3-24　设置 Android Studio 的默认字体

4. 关闭更新

单击 Settings 属性界面左侧目录中的 Appearance 选项,在右侧相关设置中取消选中 Theme 下的复选框,如图 3-25 所示。

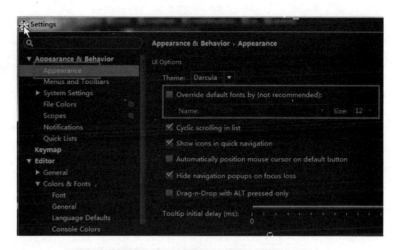

图 3-25 关闭软件的自动更新

5. 快捷键的修改

选择 File→Settings 命令,在弹出的界面中选择 Keymap 设置快捷键即可。

6. 添加 API 文档悬浮提示

Android Studio 软件默认是没有 API 文档悬浮提示的,只有按住【Ctrl + Q】组合键才会出现提示。如果要添加 API 的自动悬浮提示,可在 Settings 界面的左侧选择 Keymap,在 Keymaps 中选择 Eclipse 即可,如图 3-26 所示。

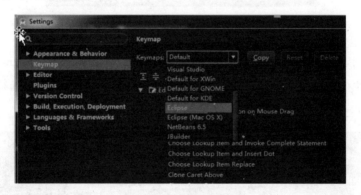

图 3-26 设置默认悬浮

7. 配置代码的自动提示

(1) 代码自动补齐。新版的 Android Studio 默认具有代码自动补齐功能(1.0 本的 Android Studio 没有补齐功能),在 Settings 界面选择 Editor 下 General 选项中的 Code Completion,选中 Autopopup code completion 复选框,自动补齐功能就设置完成,如图 3-27 所示。

(2) 代码提示快捷键。在 Settings 界面中选择左侧的 Keymap 设置快捷键编号,如图 3-28 所示。

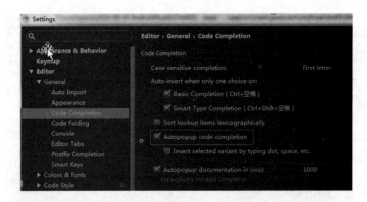

图 3-27 设置代码自动提示

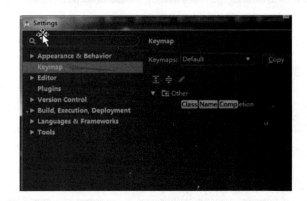

图 3-28 设置代码提示快捷键

（3）代码提示快捷键修改。在图 3-28 的搜索框输入 Android Studios name completion，就会看到代码提示的默认快捷键【Ctrl + Alt + 空格】，如果想修改快捷键，可选择 Other→Class→Name→Completion→Add Keyboard Shortcut 命令，如图 3-29 所示。

图 3-29 设置 Completion 快捷键

在图 3-29 中，右击框选部分，在弹出的对话框中进行修改。

注意：如果习惯了用 Eclipse，要注意 Android Studio 中的【Ctrl + Alt + 空格】是另外一个快捷键，单击 Keymap，在弹出的列表中选择 Cyclic Expand Word 命令即可，如图 3-30 所示。

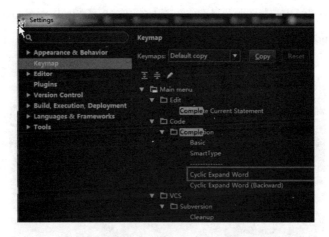

图 3-30 设置 Eclipse 快捷键

(4)配置代码提示区分大小写。Android Studio 默认的代码提示对大小写敏感。例如,输入小写的 intent,提示效果如图 3-31 所示。

图 3-31 设置代码对大小写敏感

当输入大写的 Intent 时,提示效果(这个时候出现了 Intent 类)如图 3-32 所示。

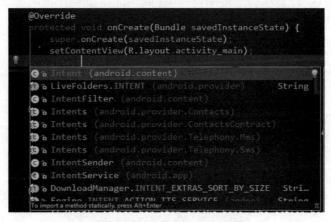

图 3-32 代码对大小写敏感

如果想让 Android Studio 对大小写不敏感,具体操作如下:进入 Settings 界面,选择 Editor 下 General 选项中的 Code Completion,在 Case sensitive completion 选项后选择 None 即可,如图 3-33 所示。

图 3-33　设置 Android Studio 代码对大小写不敏感

修改完成后,演示效果如图 3-34 所示。

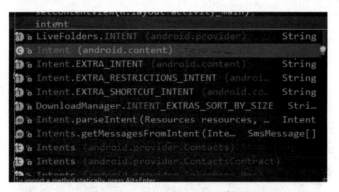

图 3-34　代码对大小写不敏感

(5)用鼠标查看源码。按住【Ctrl】键的同时单击鼠标查看源代码:(2.0 版本 Android Studio 已默认具有该设置)如果已经成功加载 SDK,其实是在快捷键中设置的,具体如下:进入 Settings 界面,选择 Keymap 选项,在右边列表框中选择 Editor Actions 即可,如图 3-35 所示。

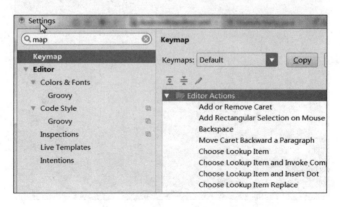

图 3-35　单击鼠标查看源码

8. 设置自动导包

在 Eclipse 中，只有每次引用一些类时必须要导入包，而 Android Studio 中则可以设置成自动导包。操作步骤如下：在 Settings 界面中，选择 Editor→General→Auto Import 选项，在界面右侧选中 Optimize imports on the fly 和 Add unambiguous imports on the fly 两个复选框，如图 3-36 所示。

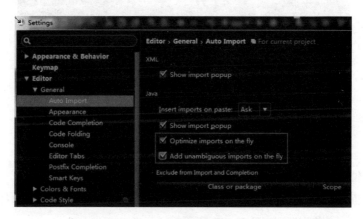

图 3-36 自动导入包

9. 显示代码行数

在 Settings 界面中，选择 Editor→General→Appearance 选项，在界面的右侧选择 Show line numbers 复选框，如图 3-37 所示。

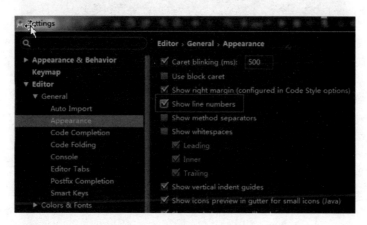

图 3-37 设置显示代码行数

10. 禁止代码折叠

IntelliJ IDEA 默认的代码基本都会自动折叠，如要禁止代码折叠，设置如下：在 Settings 界面中，选择 Editor→General→Code Folding 选项，在界面的右侧选择 One-line methods 复选框，如图 3-38所示。

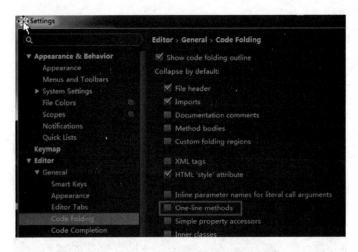

图 3-38　设置禁止代码折叠

11. 修改注释位置

在 Settings 界面中，选择 Editor→Code Style→Java 选项，在界面右侧选择设置注释，选中 Comment at first column 复选框，如图 3-39 所示。

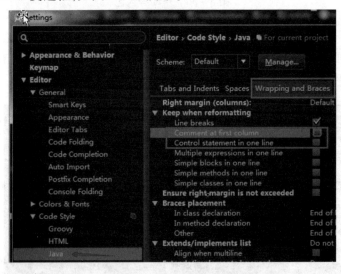

图 3-39　修改注释位置

（1）Comment at frist column：选中此项，注释的位置就会处于行首，否则会根据缩进来注释。

（2）Control statement in one line：格式化代码时，会把些很短的语句合并成一行。这样影响代码可读性，可以进行取消操作。

12. 修改新建文件文件头

在 Settings 界面中，选择 Editor→Code Style→java 选项，在界面右侧选择 Includes，在 File Header 中编辑相关的开头信息，如图 3-40 所示。

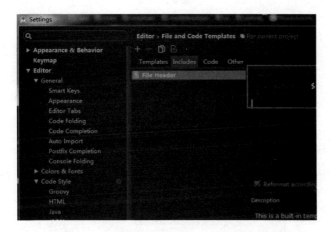

图 3-40 设置程序的开头

13. 修改文件编码为 UTF-8

Android Studio 1.1 默认的编码方式是 UTF-8，Android Studio 1.2 默认的编码方式是 GBK，Android Studio 2.1 默认的部分编码方式是 UTF-8，为了防止出错，建议统一设置为 UTF-8，在 Settings 界面中，选择 Editor→Code Style→Java 选项，在界面右侧选择 Project Encoding，选中 UTF-8 即可，如图 3-41 所示。

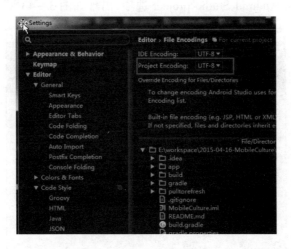

图 3-41 设置文件编码 UTF-8

3.2 创建第一个 Android 程序

3.2.1 创建第一个 Android 项目

操作步骤如下：

（1）运行 Android Studio，出现如图 3-42 所示界面。

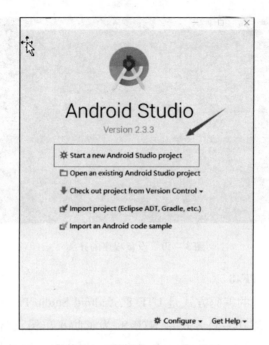

图 3-42　运行后首次出现的界面

(2) 选择 Start a new Android Studio project 选项，出现如图 3-43 所示界面。

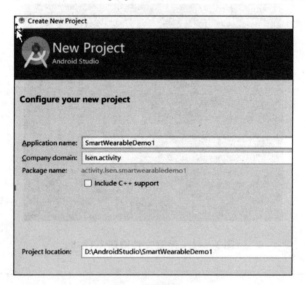

图 3-43　创建项目界面

- Application name：输入的项目名称。
- Company domain：输入公司域名或开发者简称。
- Package name：APP 打包名称。
- Project location：项目存放路径。

(3)输入完成后,单击 Next 按钮进入下一步,出现如图 3-44 所示界面。

选择设置兼容的 Android 的最低版本,依情况而定,因为蓝牙 4.0 要求 Android 系统最低版本为 4.3,所以这里选择 API 18:Android 4.3。

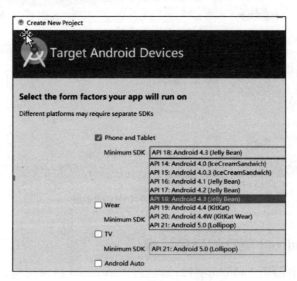

图 3-44 选择 Android 的版本

(4)选择好 Android 兼容版本后,单击 Next 按钮进入下一步,出现如图 3-45 所示界面。

选择项目的 Activity 类型,Android Studio 中会提供多种现成的模板供开发者使用,这里选择 Empty Activity。

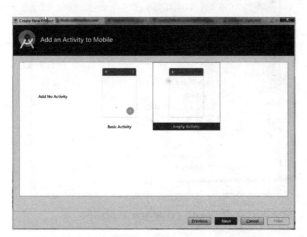

图 3-45 选择项目的 Activity 类型

(5)单击 Next 按钮进入下一步,出现如图 3-46 所示界面。

在 Activity Name 中输入 Activity 的名称,在 Layout Name 中输入界面布局的名称,最后单击 Finish 按钮完成项目的创建。

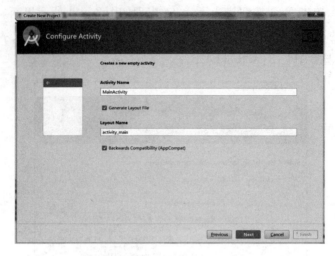

图 3-46　输入 Activity 的名称

3.2.2　熟悉 Android 项目的目录结构

在 Android Studio 中创建 Android 项目，默认进入 Android 目录结构，如图 3-47 所示。

图 3-47　Android 目录结构

一般常用的有以下两种结构：

（1）app/manifests AndroidManifest.xml：配置文件目录。

双击打开 AndroidMainifest.xml 配置文件，如图 3-48 所示。

AndroidMainifest.xml 解释如下：

● app/java：源码目录，包含工程和新建产生的 Test 工程源代码。

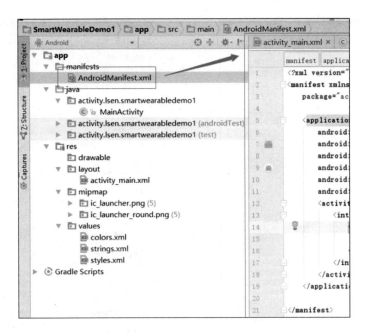

图 3-48 打开 AndroidMainifest.xml 配置文件

- app/res：资源文件目录。
- drawable：存放各种位图文件(PNG、JPEG 或 GIF)，除此之外可能是一些其他的 drawable 类型的 XML 文件。
- layout：放置 activity 布局文件。
- mipmap：应用程序启动器图标。
- colors.xml：定义颜色资源。
- strings.xml：定义字符串资源。
- styles.xml：定义样式资源。

(2) Gradle Scripts：编译相关的脚本。

3.2.3 在平板电脑中运行 Android APP

操作步骤如下：

(1) 拿出可穿戴设备实验箱中配置的平板电脑，进入设置，看是否显示"开发者选项"，如果没有则进入"关于平板电脑"中。

(2) 在"关于平板电脑"中，点击设备信息，出现如图 3-49 所示界面。

(3) 点击 5 次"版本号"，Android 系统会提示进入开发者模式。接着返回"设置"中，点击"开发者选项"进入开发者选择界面，出现如图 3-50 所示界面。

(4) 在开发者选项中，点击"开启"并选中"USB 调试"。

(5) 使用 USB 线连接 PC 与平板电脑，如图 3-51 所示。

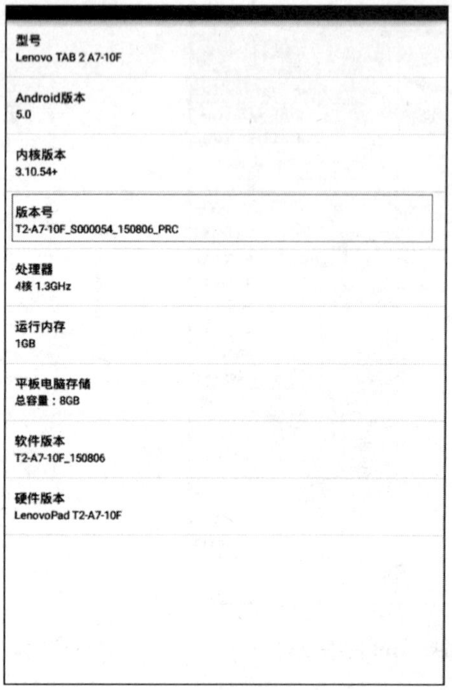

图 3-49　开发者选项内容 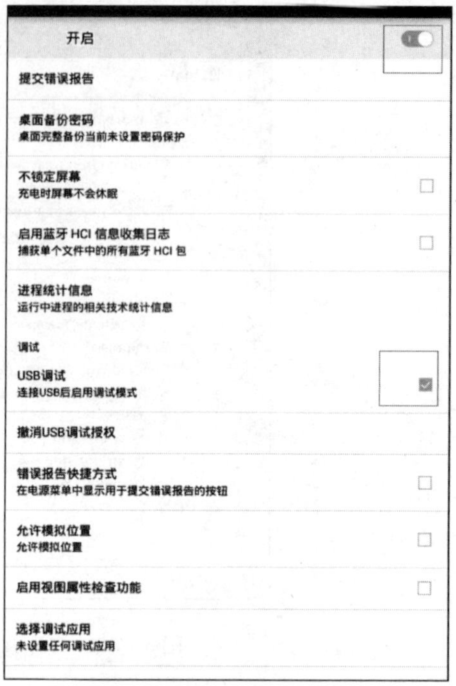　　　　　　　图 3-50　开发者选择界面

图 3-51　连接 PC 与平板电脑

(6) 连接后,便可在 Android Studio 中显示连接的 Android 设备,如图 3-52 所示。

(7) 在 Android Studio 中运行 Android 程序,在 Android 目录结构中,双击 res/layout 下的 activity_main.xml,在右侧窗口中单击 OK 按钮,如图 3-53 所示。

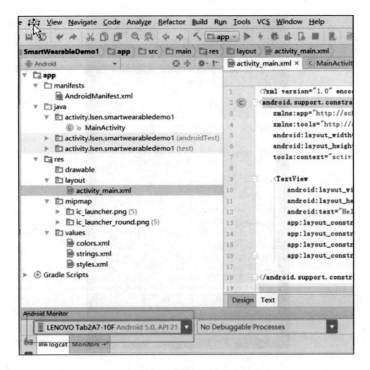

图 3-52 显示已经连接的 Android 设备

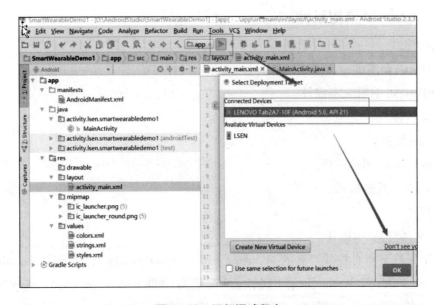

图 3-53 运行调试程序

(8) 程序安装完成后,会自动在平板电脑中显示如图 3-54 所示的信息,说明运行成功。

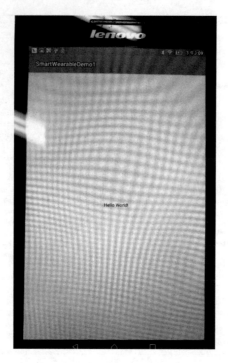

图 3-54　程序运行成功

3.2.4　Android 常用组件

1. 文本类控件

(1) TextView：负责展示文本，非编辑。

(2) EditText：可编辑文本控件。

2. 按钮类控件

(1) Button：按钮。

(2) ImageButton：图片按钮。

(3) CheckBox：复选按钮。

3. 图片控件

ImageView：负责显示图片。

以下 4 个属性是所有控件都有的 id、layout_width、layout_height 及 android:visibility。

(1) layout_width 及 layout_height 属性：可选值有两种 match_parent 和 wrap_content（从 Android 2.2 开始 fill_parent 改名为 match_parent，可以直接用 match_parent 来代替 fill_parent）。

● match_parent 表示让当前控件的大小和父布局的大小一样，也就是由父布局来决定当前控件的大小。

● wrap_content 表示让当前控件的大小能够刚好包含里面的内容，也就是由控件内容决定当前控件的大小。

（2）android：visibility 表示控件的可见属性，所有的 Android 控件都具有这个属性，可以通过 android：visibility 进行指定，可选值有 visible、invisible 和 gone 三种。

● visible 表示控件是可见的，这个值是默认值，不指定 android：visibility 时，控件都是可见的。

● invisible 表示控件不可见，但是它仍然占据着原来的位置和大小，可以理解成控件变成透明状态。

● gone 表示控件不仅不可见，而且不再占用任何屏幕空间。一般用在 Activity 中通过 setVisibility 方法来指定呈现与否。

在 Android Studio 中打开布局文件，在 Android 目录结构中，res/layout 下双击打开 activity_main.xml 文件，在代码中可以直接编辑相关控件，如图 3－55 所示。

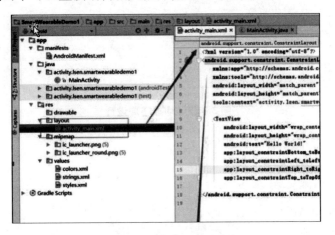

图 3－55　studio 的界面布局

布局文件默认打开 Text，可以手动输入代码进行布局设计，如图 3－56 所示。

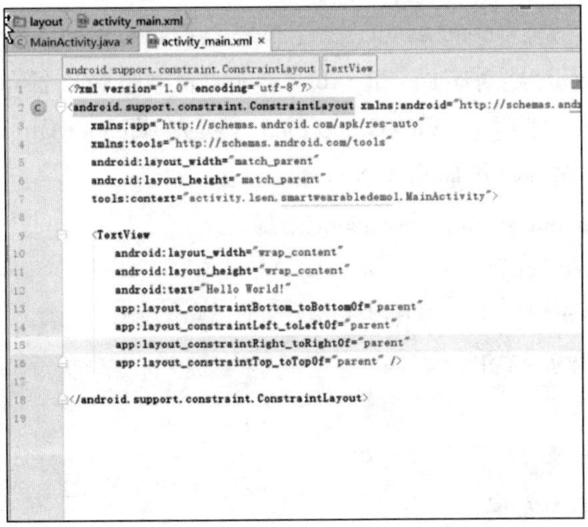

图 3－56　手动输入代码进行布局设计

单击 App Theme 切换布局设计,如图3-57所示。

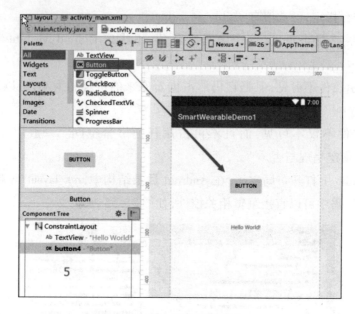

图3-57 切换布局设计

在 Design 中拖动控件到布局中进行设计。

(1)方框1:界面横竖屏切换。

(2)方框2:预览的设备。

(3)方框3:Android API。

(4)方框4:主题。

(5)方框5:组件的结构,点击可快速定位到相应组件。

3.2.5 Android 常用布局

常用的五大布局分别是:LinearLayout(线性布局)、FrameLayout(框架布局)、RelativeLayout(相对布局)、AbsoluteLayout(绝对布局)和 TableLayout(表格布局),各布局之间可以互相嵌套。

所有的布局方式都可以归类为 ViewGroup 的 5 个类别,即 ViewGroup 的 5 个直接子类(5 种布局方式)。其他的一些布局都扩展自这 5 个类。

本节将介绍 LinearLayout 和 RelativeLayout 布局。

1. LinearLayout(线性布局)

在一个方向(垂直或水平)对齐所有子元素。

android:orientation = "vertical"

(1)vertical:垂直对齐。

(2)horizontal:水平对齐。

2. RelativeLayout(相对布局)

相对布局的子控件会根据它们所设置的参照控件和参数进行相对布局。参照控件可以是父控件,也可以是其他子控件,但是被参照的控件必须要在参照它的控件之前定义。

(1) android:layout_above:将该控件的底部置于给定 ID 的控件之上。

(2) android:layout_below:将该控件的底部置于给定 ID 的控件之下。

(3) android:layout_toLeftOf:将该控件的右边缘与给定 ID 的控件左边缘对齐。

(4) android:layout_toRightOf:将该控件的左边缘与给定 ID 的控件右边缘对齐。

(5) android:layout_centerHorizontal:如果为 true,将该控件的右边缘与给定 ID 置于水平居中。

(6) android:layout_centerVertical:如果为 true,将该控件的右边缘与给定 ID 置于垂直居中。

(7) android:layout_centerInParent:如果为 true,将该控件的右边缘与给定 ID 置于父控件的中央。

3.3 Android 用户登录及注册实验

3.3.1 登录界面 UI 设计

Activity_login.xml 关键部分代码如下,其他代码请查看项目文件包。

1. 用户名输入栏

```xml
<LinearLayout              //用户名输入框布局
   android:layout_width = "match_parent"
   android:layout_height = "@dimen/line_height"
   android:layout_marginLeft = "@dimen/left_right"
   android:layout_marginRight = "@dimen/left_right"
   android:background = "@drawable/underline_shape"
   android:gravity = "center"
   android:orientation = "horizontal"
   android:paddingLeft = "@dimen/left_right"
   android:paddingRight = "@dimen/left_right" >
   <TextView              //文本编辑框
      android:layout_width = "@dimen/line_height2"
      android:layout_height = "match_parent"
      android:gravity = "center_vertical"
      android:text = "@string/login_name"
      android:textColor = "@color/text_color"
      android:textSize = "@dimen/text_size" />
```

```xml
<EditText                  //输入内容信息
    android:id = "@+id/login_nameEt"
    android:layout_width = "0dp"
    android:layout_height = "match_parent"
    android:layout_weight = "1"
    android:background = "@null"
    android:ems = "10"
    android:hint = "@string/hint_login_name"
    android:textColor = "@color/text_color" />
```

2. 密码输入栏

```xml
<LinearLayout              // 密码输入框布局
    android:layout_width = "match_parent"
    android:layout_height = "@dimen/line_height"
    android:layout_marginLeft = "@dimen/left_right"
    android:layout_marginRight = "@dimen/left_right"
    android:layout_marginTop = "@dimen/left_right"
    android:background = "@drawable/underline_shape"
    android:gravity = "center"
    android:orientation = "horizontal"
    android:paddingLeft = "@dimen/left_right"
    android:paddingRight = "@dimen/left_right" >
<TextView                  //设置密码编辑框1
    android:layout_width = "@dimen/line_height2"
    android:layout_height = "match_parent"
    android:gravity = "center_vertical"
    android:text = "@string/login_pAndroid Studiosword"
    android:textColor = "@color/text_color"
    android:textSize = "@dimen/text_size" />
<EditText                  //设置密码编辑框2
    android:id = "@+id/login_pAndroid StudioswordEt"
    android:layout_width = "0dp"
    android:layout_height = "match_parent"
    android:layout_weight = "1"
```

```
android:background = "@null"
android:ems = "10"
android:hint = "@string/hint_login_pAndroid Studiosword"
android:inputType = "textPAndroid Studiosword"
android:textColor = "@color/text_color" />
```

3. 登录栏

```
<TextView
    android:id = "@+id/loginTv"
    android:layout_width = "match_parent"
    android:layout_height = "@dimen/line_height"
    android:layout_marginLeft = "@dimen/left_right"
    android:layout_marginRight = "@dimen/left_right"
    android:layout_marginTop = "@dimen/left_right"
    android:background = "@color/text_color"
    android:gravity = "center"
    android:text = "@string/login"
    android:textColor = "@color/white"
    android:textSize = "@dimen/text_size" />
```

4. 免登录及用户注册栏

```
<LinearLayout              //设置布局
    android:layout_width = "match_parent"
    android:layout_height = "@dimen/line_height"
    android:layout_marginLeft = "@dimen/left_right"
    android:layout_marginRight = "@dimen/left_right"
    android:layout_marginTop = "@dimen/left_right" >
<TextView                  //设置登录用户名
    android:id = "@+id/login_exemptTv"
    android:layout_width = "0dp"
    android:layout_height = "match_parent"
    android:layout_weight = "1"
    android:gravity = "center|start"
    android:text = "@string/exempt_login"
    android:textColor = "@color/text_color"
    android:textSize = "@dimen/text_size" />
```

```
<TextView                //设置密码
    android:id = "@+id/login_registerTv"
    android:layout_width = "0dp"
    android:layout_height = "match_parent"
    android:layout_weight = "1"
    android:gravity = "center|end"
    android:text = "@string/user_register"
    android:textColor = "@color/text_color"
    android:textSize = "@dimen/text_size" />
</LinearLayout>
```

登录界面 UI 布局如图 3 – 58 所示。

图 3 – 58　登录界面 UI

3.3.2　注册界面 UI 设计

Activity_register.xml 关键部分代码如下,其他代码请查看项目文件包。

1. 用户名输入栏

```
<LinearLayout
    android:layout_width = "match_parent"
    android:layout_height = "@dimen/line_height"
    android:layout_marginLeft = "@dimen/left_right"
    android:layout_marginRight = "@dimen/left_right"
    android:background = "@drawable/underline_shape"
    android:gravity = "center"
    android:orientation = "horizontal"
    android:paddingLeft = "@dimen/left_right"
    android:paddingRight = "@dimen/left_right" >
    <TextView
        android:layout_width = "@dimen/line_height2"
        android:layout_height = "match_parent"
        android:gravity = "center_vertical"
        android:text = "@string/login_name"
        android:textColor = "@color/text_color"
        android:textSize = "@dimen/text_size" />
```

```xml
<EditText
    android:id = "@+id/register_nameEt"
    android:layout_width = "0dp"
    android:layout_height = "match_parent"
    android:layout_weight = "1"
    android:background = "@null"
    android:ems = "10"
    android:hint = "@string/hint_login_name" />
</LinearLayout>
```

2. 密码输入栏

```xml
<LinearLayout
    android:layout_width = "match_parent"
    android:layout_height = "@dimen/line_height"
    android:layout_marginLeft = "@dimen/left_right"
    android:layout_marginRight = "@dimen/left_right"
    android:layout_marginTop = "@dimen/left_right"
    android:background = "@drawable/underline_shape"
    android:gravity = "center"
    android:orientation = "horizontal"
    android:paddingLeft = "@dimen/left_right"
    android:paddingRight = "@dimen/left_right" >
<TextView            //密码输入
    android:layout_width = "@dimen/line_height2"
    android:layout_height = "match_parent"
    android:gravity = "center_vertical"
    android:text = "@string/login_pAndroid Studiosword"
    android:textColor = "@color/text_color"
    android:textSize = "@dimen/text_size" />
<EditText            //密码编辑框
        android:id = "@+id/register_pAndroid StudioswordEt"
        android:layout_width = "0dp"
        android:layout_height = "match_parent"
        android:layout_weight = "1"
        android:background = "@null"
        android:ems = "10"
        android:hint = "@string/hint_login_pAndroid Studiosword"
        android:inputType = "textPAndroid Studiosword" />
</LinearLayout>
```

3. 确认密码输入栏

```xml
<LinearLayout
    android:layout_width="match_parent"
    android:layout_height="@dimen/line_height"
    android:layout_marginLeft="@dimen/left_right"
    android:layout_marginRight="@dimen/left_right"
    android:layout_marginTop="@dimen/left_right"
    android:background="@drawable/underline_shape"
    android:gravity="center"
    android:orientation="horizontal"
    android:paddingLeft="@dimen/left_right"
    android:paddingRight="@dimen/left_right" >
<TextView            //文本编辑框
    android:layout_width="@dimen/line_height2"
    android:layout_height="match_parent"
    android:gravity="center_vertical"
    android:text="@string/confirm_pAndroid Studiosword"
    android:textColor="@color/text_color"
    android:textSize="@dimen/text_size" />
<EditText            //用户名编辑框
    android:id="@+id/register_confirm_pAndroid StudioswordEt"
    android:layout_width="0dp"
    android:layout_height="match_parent"
    android:layout_weight="1"
    android:background="@null"
    android:ems="10"
    android:hint="@string/hint_login_pAndroid Studiosword"
    android:inputType="textPAndroid Studiosword" />
</LinearLayout>
```

4. 注册提交栏

```xml
<TextView
    android:id="@+id/registerTv"
    android:layout_width="match_parent"
    android:layout_height="@dimen/line_height"
```

```
android:layout_marginLeft = "@dimen/left_right"
android:layout_marginRight = "@dimen/left_right"
android:layout_marginTop = "@dimen/left_right"
android:background = "@color/text_color"
android:gravity = "center"
android:text = "@string/register"
android:textColor = "@color/white"
android:textSize = "@dimen/text_size" / >
```

3.3.3　注解控件 ButterKnife 的使用

ButterKnife 大大简化了程序员编写代码的速度,并且在 7.0 版本以后引入了注解处理器,取代了之前利用反射原理进行 findViewById(提取已经写好的 View 对象)影响 APP 性能的方式,不再影响 APP 运行效率。

ButterKnife 安装方法:按住【Ctrl + Alt + S】组合键打开 Settings 界面,在搜索框中输入 ButterKnife 搜索该 ButterKnife 插件,单击 Install plugin 按钮安装即可,如图 3 – 59 所示。

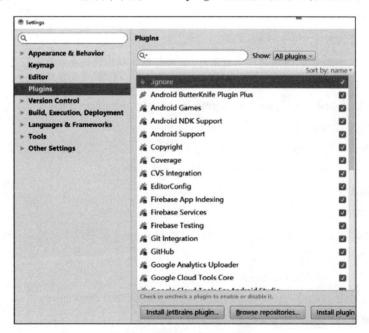

图 3 – 59　ButterKnife 安装

3.3.4　实现 Android 与服务器间的 HTTP 通信

(1)HTTP 通信通过传递实体对象进行数据通信。实现代码如下:

```java
//http 通信类 HttpSend.java
public static Object SendObject(String urlStr, Object object) {
    URL url = null;
    Object result = null;//要返回的结果
    try {
        url = new URL(urlStr);
        HttpURLConnection httpURLConnection = (HttpURLConnection) url.
            openConnection();
        httpURLConnection.setConnectTimeout(2000);//设置连接超时时间,单位 ms
        httpURLConnection.setReadTimeout(2000);    //设置读取超时时间,单位 ms
        //设置是否向 httpURLConnection 输出,因为 post 请求参数要放在 http 正文内,所以要设
        //置为 true
        httpURLConnection.setDoOutput(true);
        //设置是否从 httpURLConnection 读入,默认是 false
        httpURLConnection.setDoInput(true);
        //POST 请求不能用缓存,设置为 false
        httpURLConnection.setUseCaches(false);
        //传送的内容是可序列化的
        //如果不设置此项,传送序列化对象时,当 Web 服务默认的不是这种类型时,会抛出 java.io.
        //EOFException 错误
        httpURLConnection.setRequestProperty("Content-type",
            "application/x-java-serialized-object");
        //设置请求方法是 POST
        httpURLConnection.setRequestMethod("GET");
        //连接服务器
        httpURLConnection.connect();
        //getOutputStream 会隐含调用 connect(),所以不用写上述的 httpURLConnection.
        //connect()也行
        //得到 httpURLConnection 的输出流
        OutputStream os = httpURLConnection.getOutputStream();
        //构建输出流对象,以实现输出序列化的对象
        ObjectOutputStream objOut = new ObjectOutputStream(os);
        //向对象输出流写出数据,这些数据将存到内存缓冲区中
        objOut.writeObject(object);
        //刷新对象输出流,将字节全部写入输出流中
```

```
            objOut.flush();
            //关闭流对象
            objOut.close();
            os.close();
            //将内存缓冲区中封装好的完整的 HTTP 请求电文发送到服务端,并获取访问状态
            if (HttpURLConnection.HTTP_OK == httpURLConnection.getResponseCode()) {
                //得到 httpURLConnection 的输入流,这里包含服务器返回来的 Java 对象
                InputStream in = httpURLConnection.getInputStream();
                //构建对象输入流,使用 readObject()方法取出输入流中的 Java 对象
                ObjectInputStream inObj = new ObjectInputStream(in);
                object = inObj.readObject();
                //取出对象里面的数据
                result = object;
                //输出日志,在控制台可以看到接收到的数据
                Log.w("HTTP", result + ":by post");
                //关闭创建的流
                in.close();
                inObj.close();
            } else {
                result = new Object();
            }
        } catch (Exception e) {
            e.printStackTrace();
        }
        return result;
    }
}
```

（2）LoginActivity 登录。实现代码如下：

```
//获得登录界面组件的应用
public clAndroid Studios LoginActivity extends AppCompatActivity {
    //获得登录界面组件
    @BindView(R.id.login_nameEt)
    EditText loginNameEt;
    @BindView(R.id.login_pAndroid StudioswordEt)
    EditText loginPAndroid StudioswordEt;
```

```java
@BindView(R.id.loginTv)
TextView loginTv;
@BindView(R.id.login_exemptTv)
TextView loginExemptTv;
@BindView(R.id.login_registerTv)
TextView loginRegisterTv;
User user;              //定义一个用户变量
private Vibrator vibrator;
@Override
protected void onCreate(@Nullable Bundle savedInstanceState) {
    super.onCreate(savedInstanceState);
    setRequestedOrientation(ActivityInfo.SCREEN_ORIENTATION_PORTRAIT);
    setContentView(R.layout.activity_login);    //设置进入登录界面
    ButterKnife.bind(this);
    vibrator = (Vibrator) getSystemService(Context.VIBRATOR_SERVICE);
}
```

(3) 通过 URL 和实例化对象 User 进行用户名和密码的数据通信。实现代码如下：

```java
@OnClick({R.id.loginTv, R.id.login_exemptTv, R.id.login_registerTv})
public void onViewClicked(View view) {
    Intent intent;
    switch (view.getId()) {      //判断点击了哪个按钮
        cAndroid Studioe R.id.loginTv:
            vibrator.vibrate(100);
            if(isCheck()) {
                User user = new User();
                user.setType("login");
                user.setUserName(loginNameEt.getText().toString());
                user.setUserPAndroid Studiosword(loginPAndroid StudioswordEt.
                    getText().toString());
                login(user);
            }
            break;
        cAndroid Studioe R.id.login_exemptTv:
```

```
            vibrator.vibrate(100);              //设置间隔时间
            intent = new Intent(LoginActivity.this, MainActivity.clAndroid Studios);
            startActivity(intent);
            finish();                            //关闭程序,停止
            break;
        cAndroid Studioe R.id.login_registerTv:
            vibrator.vibrate(100);
            intent = new Intent(LoginActivity.this, RegisterActivity.clAndroid
                Studios);
            startActivity(intent);               //跳转到新的界面
            break;
}
```

登录返回成功则跳转到 MainActivity 主界面,失败则出现提示信息。实现代码如下:

```
private void login(final User use) {
    LoadingDialog.show(LoginActivity.this, "请等待...");
    new Thread(new Runnable() {
        @Override
        public void run() {
            String url = "http://192.168.1.21:8092/WearableDevice/DatAndroid
                Studioervlet";                    //设置服务器网址
            Object o = HttpSend.SendObject(url, use);
            if (o != null) {
                user = (User) o;
                if (user.getType().equals("success")) {  //判断是否连接成功
                    LoadingDialog.dismiss(LoginActivity.this);
                    Intent intent = new Intent(LoginActivity.this, MainActivity.
                        clAndroid Studios);
                    Bundle bundle = new Bundle();
                    bundle.putSerializable("User", user);
                    intent.putExtrAndroid Studio(bundle);
                    startActivity(intent);
                    finish();
                } else {                          //判断消息类型
                    LoadingDialog.dismiss(LoginActivity.this);
                    showToAndroid Studiot("用户名或密码错误");
                }
```

```
                } else {                    //判断消息类型
                    LoadingDialog.dismiss(LoginActivity.this);
                    showToAndroid Studiot("无法连接服务器");
                }
            }
    }).start();
}
```

在 AndroidManifest 中加入 Android 网络访问权限：

```
<uses-permission android:name="android.permission.INTERNET"/>
```

3.4　Android 程序签名打包

1. 签名的概念和作用

Android APP 需要用一个证书对应用进行数字签名，否则将无法安装到 Android 手机上，平时调试运行时，Android Studio 会自动用默认的密钥和证书进行签名；但是，实际发布编译时，则不会自动签名，这时就需要进行手动签名。

2. 为 Android 安装包(APK)签名的优点

（1）应用程序升级：如果希望用户无缝升级到新的版本，就必须用同一个证书进行签名。这是由于只有以同一个证书签名，系统才会允许安装升级应用程序。如果采用了不同的证书，系统就会要求应用程序采用不同的包名称，在这种情况下相当于安装了一个全新的应用程序。如果想升级应用程序，签名证书要相同，包名称要相同。

（2）应用程序模块化：Android 系统可以允许同一个证书签名的多个应用程序在一个进程中运行，系统实际把他们作为一个单个的应用程序，此时就可以把应用程序以模块的方式进行部署，而用户可以独立升级其中的一个模块。

（3）代码或者数据共享：Android 提供了基于签名的权限机制，那么一个应用程序就可以为另一个以相同证书签名的应用程序公开自己的功能。以同一个证书对多个应用程序进行签名，利用基于签名的权限检查，就可以在应用程序间以安全的方式共享代码和数据。不同的应用程序之间，想共享数据，或者共享代码，就要让它们运行在同一个进程中，而且要让它们用相同的证书签名。

3. Android Studio 如何打包签名

因为学习本课程的都是初学者，多渠道打包的内容以后再进行讲解。本节只讲最简单的打包调试时默认生成的 apk，在 app/build/outputs/apk 目录下，和 Eclipse 并不相同。Eclipse 是在 bin 目录下生成的。

打开 Android Studio 中的 Hello World 项目然后进行以下操作：

(1) 选择 Build→Generate Signed APK 命令，如图 3-60 所示。

(2) 在弹出的窗口中创建一个 Key，单击 Next 按钮继续，如图 3-61 所示。

图 3-60　选择 Generate Signed APK 命令

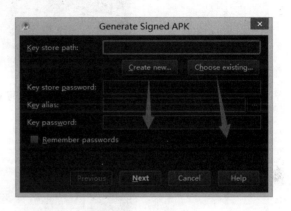

图 3-61　创建新 key

(3) 根据自己的需要填写相关信息如图 3-62 所示。

图 3-62　填写签名的相关信息

(4) 单击 OK 按钮后，在弹出的对话框中填写密码信息，如图 3-63 所示。

图 3 – 63　填写 Key 的密码信息

（5）单击 Next 按钮，完成创建签名，如图 3 – 64 所示。

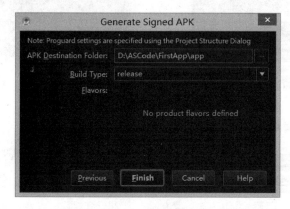

图 3 – 64　完成创建签名

（6）单击 Finish 按钮，稍等一会儿，打包签名成功后，即可看到打包后的 APK 已经出现在 app 目录下，如图 3 – 65 所示。

图 3 – 65　打包后 APK 出现的目录

打包 Android APK 的方法有很多种，如命令行，或者 Gradle、ANT、MAVEN 等，本节讲解了最简单的通过图形化界面打包签名的方式。

3.5 Android TextView(文本框)详解

本节讲解的 TextView(文本框)是用于显示文本的一个控件。首先介绍以下几个单位:

(1) dp(dip):device independent pixels(设备独立像素),不同设备有不同的显示效果,与设备硬件有关,一般为了支持 WVGA、HVGA 和 QVGA 推荐使用此单位。

(2) px:pixels(像素),不同设备显示效果相同,一般 HVGA 代表 320×480 像素,用得比较多。

(3) pt:point,是一个标准的长度单位,1pt = 1/72in,用于印刷业,非常简单易用。

(4) sp:scaled pixels(放大像素),主要用于字体显示。

1. 基础属性详解

通过下面这个简单的界面了解几个最基本的属性,图 3 - 66 所示为常见的文本框。

图 3 - 66 常见的文本框

实现代码如下:

```
<RelativeLayout xmlns:android = "http://schemAndroid Studio.android.com/apk/res/
    android"
  xmlns:tools = "http://schemAndroid Studio.android.com/tools"
  android:layout_width = "match_parent"      //设置布局方式
  android:layout_height = "match_parent"
  tools:context = ".MainActivity"              //设置函数名称
  android:gravity = "center"
  android:background = "#8fffad" >
```

```xml
<TextView
    android:id="@+id/txtOne"
    android:layout_width="200dp"        //设置大小
    android:layout_height="200dp"
    android:gravity="center"
    android:text="TextView"              //显示框
    android:textColor="#EA5246"
    android:textStyle="bold|italic"
    android:background="#000000"        //设置背景颜色
    android:textSize="18sp" />
</RelativeLayout>
```

上面的 TextView 中有下述几个属性：

(1) id：为 TextView 设置一个组件 id，据此可以在 Java 代码中通过 findViewById() 方法获取该对象，然后进行相关属性的设置。使用 RelativeLayout 时，参考组件用的也是 id。

(2) layout_width：组件的宽度，一般写 ** wrap_content ** 或者 ** match_parent(fill_parent) **，前者是控件显示的内容多大，控件就多大，而后者会填满该控件所在的父容器；当然，也可以设置成特定的大小，例如，这里为了显示效果，设置成了 200 dp。

(3) layout_height：组件的高度，设置方法同上。

(4) gravity：设置控件中内容的对齐方向，TextView 中是文字，ImageView 中是图片。

(5) text：设置显示的文本内容，一般是把字符串写到 string.xml 文件中，然后通过 @String/xxx 取得对应的字符串内容。

(6) textColor：设置字体颜色，通过 colors.xml 资源来引用。

(7) textStyle：设置字体风格，有 ** normal **（无效果）、** bold **（加粗）、** italic **（斜体）3 个可选值。

(8) textSize：字体大小，单位一般用 sp。

(9) background：控件的背景颜色，可以理解为填充整个控件的颜色，可以是图片。

2. 带阴影的 TextView

涉及的几个属性：

(1) android:shadowColor——设置阴影颜色，需要与 shadowRadius 一起使用。

(2) android:shadowRadius——设置阴影的模糊程度，设为 0.1 变成字体颜色，建议使用 3.0。

(3) android:shadowDx——设置阴影在水平方向的偏移，即水平方向阴影开始的横坐标位置。

(4) android:shadowDy——设置阴影在竖直方向的偏移，即竖直方向阴影开始的纵坐标位置。

带阴影的文本框效果如图 3-67 所示。

图 3-67 带阴影的文本框效果

实现代码如下：

```
<TextView
    android:layout_width = "wrap_content"
    android:layout_height = "wrap_content"
    android:layout_centerInParent = "true"
    android:shadowColor = "#F9F900"
    android:shadowDx = "10.0"
    android:shadowDy = "10.0"
    android:shadowRadius = "3.0"
    android:text = "带阴影的 TextView"
    android:textColor = "#4A4AFF"
    android:textSize = "30sp" />
```

3. 带边框的 TextView

如果你想为 TextView 设置一个边框背景,可以采用普通矩形边框或者圆角边框。另外,TextView 是很多其他控件的父类,如 Button。也可以自行编写一个 ShapeDrawable 的资源文件,然后在 TextView 中将 blackgroung 设置为这种资源即可。

简单介绍一下 ShapeDrawable 资源文件的几个结点及属性：

（1）< solid android:color = " xxx " > :设置背景颜色的属性。

（2）< stroke android:width = " xdp " android:color = " xxx " > :设置边框的粗细及边框的颜色。

（3）< padding androidLbottom = " xdp " … > :设置边距。

（4）< corners android:topLeftRadius = " 10px " … > :设置圆角。

（5）< gradient > 设置渐变色,可选属性有 startColor（起始颜色）；endColor（结束颜色）；centerColor（中间颜色）；angle（方向角度）,等于 0 时,从左到右,然后逆时针方向转。

实现效果如图 3 – 68 所示。

图 3 – 68　带边框的文本框效果

实现代码如下:

1. 编写矩形边框的 Drawable

```
< ? xml version = "1.0" encoding = "utf - 8"? >
< shape xmlns:android = "http://schemAndroid Studio.android.com/apk/res/android" >
<!-- 设置一个黑色边框 -->
< stroke android:width = "2px" android:color = "#000000"/>
<!-- 渐变 -->
< gradient
    android:angle = "270"
    android:endColor = "#C0C0C0"
    android:startColor = "#FCD209" />
<!-- 设置一下边距,让空间大一点 -->
< padding
```

```
    android:left = "5dp"
    android:top = "5dp"
    android:right = "5dp"
    android:bottom = "5dp"/>
</shape>
```

2. 编写圆角矩形边框的 Drawable

```
<?xml version = "1.0" encoding = "utf-8"?>
<shape xmlns:android = "http://schemAndroid Studio.android.com/apk/res/android">
<!-- 设置透明背景色 -->
<solid android:color = "#87CEEB"/>
<!-- 设置一个黑色边框 -->
<stroke
    android:width = "2px"
    android:color = "#000000"/>
<!-- 设置四个圆角的半径 -->
<corners
    android:bottomLeftRadius = "10px"
    android:bottomRightRadius = "10px"
    android:topLeftRadius = "10px"
    android:topRightRadius = "10px"/>
<!-- 设置一下边距,让空间大一点 -->
<padding
    android:bottom = "5dp"
    android:left = "5dp"
    android:right = "5dp"
    android:top = "5dp"/>

</shape>
```

3. 将 TextView 的 blackground 属性设置成上面这两个 Drawable

```
<LinearLayout xmlns:android = "http://schemAndroid Studio.android.com/apk/res/android"
    xmlns:tools = "http://schemAndroid Studio.android.com/tools"
```

```
        android:layout_width = "match_parent"
        android:layout_height = "match_parent"
        android:background = "#FFFFFF"
        android:gravity = "center"
        android:orientation = "vertical"
        tools:context = ".MainActivity" >
    <TextView
        android:id = "@ + id/txtOne"
        android:layout_width = "200dp"
        android:layout_height = "64dp"
        android:textSize = "18sp"
        android:gravity = "center"
        android:background = "@drawable/txt_rectborder"
        android:text = "矩形边框的 TextView" />
    <TextView
        android:id = "@ + id/txtTwo"
        android:layout_width = "200dp"
        android:layout_height = "64dp"
        android:layout_marginTop = "10dp"
        android:textSize = "18sp"
        android:gravity = "center"
        android:background = "@drawable/txt_radiuborder"
        android:text = "圆角边框的 TextView" />

</LinearLayout>
```

3.6 EditText(输入框)详解

上一节中介绍了第一个 UI 控件 TextView(文本框),文中给出了很多实际开发中可能遇到的一些需求的解决方法,给用户开发带来便利。本节将介绍第二个很常用的控件 EditText(输入框),它与 TextView 非常类似,最大的区别是 EditText 可以接收用户输入。

3.6.1 设置默认提示文本

对于用户登录界面我们并不陌生,很多时候都会用到这种界面,如图 3 – 69 所示。

这里只介绍默认提示文本的两个控制属性：

```
android:hint = "默认提示文本"
android:textColorHint = "#95A1AA"
```

前者设置提示的文本内容，后者设置提示文本的颜色。

3.6.2 获得焦点后全选组件内所有文本内容

如果在点击输入框获得焦点后，不是将光标移动到文本的开始或者结尾，而是获取到输入框中所有的文本内容，可以使用 selectAllOnFocus 属性。

```
android:selectAllOnFocus = "true"
```

图 3-69 登录界面

如图 3-70 所示，第一个设置了该属性，第二个没设置该属性，设置为 true 的 EditText 获得焦点后选中的是所有文本。

图 3-70 编辑框截图

3.6.3 限制 EditText 输入类型

有时可能需要对输入的数据进行限制，例如，输入电话号码时，如果输入了一串字母，显然是不符合预期的，而限制输入类型可以通过 inputType 属性来实现。

例如，限制只能为电话号码、密码：

```
<EditText
android:layout_width="fill_parent"
android:layout_height="wrap_content"        //布局
android:inputType="phone"/>                 //手机类型
//数值类型
android:inputType="number"                  //手机号码
android:inputType="numberSigned"
android:inputType="numberDecimal"           //出厂日期
android:inputType="phone"                   //拨号键盘
android:inputType="datetime"
android:inputType="date"                    //日期键盘
android:inputType="time"                    //时间键盘
```

3.6.4　设置最小行、最多行、单行、多行、自动换行

EditText 默认是多行显示的,并且能够自动换行,即当一行显示不完时,会自动换到第二行,如图 3-71 所示。

图 3-71　能够自动换行的编辑框

可以对其进行限制,例如,设置最小行的行数

```
android:minLines="3"
```

或者设置 EditText 最大的行数:

```
android:maxLines="3"
```

当输入内容超过 maxLine 时,文字会自动向上滚动。

另外,很多时候可能要限制 EditText 只允许单行输入,而且不会滚动。例如,上面的登录界面的例子,只需要设置

```
android:singleLine = "true"
```

即可实现单行输入不换行。

3.6.5 设置文字间隔及英文字母大写类型

可以通过下述两个属性来设置字的间距:

```
android:textScaleX = "1.5"//设置字与字的水平间隔
android:textScaleY = "1.5"//设置字与字的垂直间隔
```

另外,EditText 还提供了设置英文字母大写类型的属性 android:capitalize,默认为 none,3 个可选值如下:

(1) sentences:仅第一个字母大写。

(2) words:每一个单词首字母大写,用空格区分单词。

(3) characters:每一个英文字母都大写。

3.6.6 控制 EditText 四周的间隔距离与内部文字与边框间的距离

使用 margin 相关属性增加组件相对其他控件的距离,例如 android:marginTop = "5dp" 使用 padding 增加组件内文字和组件边框的距离,如 android:paddingTop = "5dp"。

3.6.7 设置 EditText 获得焦点,同时弹出小键盘

进入 Activity 后,让 EditText 获得焦点,同时弹出小键盘供用户输入,操作步骤如下:

(1) 让 EditText 获得焦点与清除焦点:

```
edit.requestFocus();     //请求获取焦点
edit.clearFocus();       //清除焦点
```

(2) 获得焦点后,弹出小键盘:

- 低版本的系统直接请求 requestFocus 自动弹出小键盘。
- 稍微高一点的版本则需要手动弹出键盘:

```
InputMethodManager imm =(InputMethodManager) getSystemService(Context.INPUT_
    METHOD_SERVICE);
imm.toggleSoftInput(0,InputMethodManager.HIDE_NOT_ALWAYS);
```

或者

```
InputMethodManager imm =(InputMethodManager) getSystemService(Context.INPUT_
    METHOD_SERVICE);
imm.showSoftInput(view,InputMethodManager.SHOW_FORCED);
imm.hideSoftInputFromWindow(view.getWindowToken(),0);//强制隐藏键盘
```

如果两种方法并没有弹出小键盘，可以使用 windowSoftInputMode 属性解决弹出小键盘的问题：

android:windowSoftInputMode 是 Activity 主窗口与软键盘的交互模式，可以用来避免输入法面板遮挡问题，是 Android1.5 后的一个新特性。

这个属性能影响两件事情：

（1）当有焦点产生时，软键盘是隐藏还是显示。

（2）是否减少活动主窗口大小以便腾出空间放软键盘。

用来放小键盘有下述值可供选择，可设置多个值，用"|"分开：

● stateUnspecified：软键盘的状态并没有指定，系统将选择一个合适的状态或依赖于主题的设置。

● stateUnchanged：当这个 Activity 出现时，软键盘将一直保持在上一个 Activity 里的状态，无论是隐藏还是显示。

● stateHidden：用户选择 Activity 时，软键盘总是被隐藏。

● stateAlwaysHidden：当该 Activity 主窗口获取焦点时，软键盘也总是被隐藏的。

● stateVisible：软键盘通常是可见的。

● stateAlwaysVisible：用户选择 Activity 时，软键盘总是显示的状态。

● adjustUnspecified：默认设置，通常由系统自行决定是隐藏还是显示。

● adjustResize：该 Activity 总是调整屏幕的大小以便留出软键盘的空间。

● adjustPan：当前窗口的内容将自动移动，以便当前焦点从不被键盘覆盖且用户总能看到输入内容的部分。

用户可以在 AndroidManifest.xml 为需要弹出小键盘的 Activity 设置这个属性，然后在 EditText 对象设置 requestFocus() 就可以实现效果。

3.6.8　EditText 光标位置的控制

有时可能需要控制 EditText 中的光标移动到指定位置或者选中某些文本，此时可用 EditText 提供 setSelection() 的方法；也可以调用 setSelectAllOnFocus(true) 让 EditText 获得焦点时选中全部文本。

另外，还可以调用 setCursorVisible(false) 设置光标不显示；还可以调用 getSelectionStart() 和 getSelectionEnd() 获得当前光标的前后位置。

3.6.9 带表情的 EditText 的简单实现

相信大家对于 QQ 或者微信都很熟悉,发送文本时可以连同表情一起发送。

实现方式有 2 种:使用 SpannableString 来实现;使用 HTML 类来实现。

下面用的是第一种方式,这里只实现一个简单的效果,用户可以把方法抽取出来,自定义一个 EditText。

也可以自己动手写个类似于 QQ 那样有多个表情选择的输入框,如图 3-72 所示。

图 3-72 带表情的编辑框

实现代码如下:

```
public clAndroid Studios MainActivity extends Activity{
    private Button btn_add;
    private EditText edit_one;
    @Override
    protected void onCreate(Bundle savedInstanceState){
        super.onCreate(savedInstanceState);
        setContentView(R.layout.activity_main);
        btn_add = (Button)findViewById(R.id.btn_add);
        edit_one = (EditText)findViewById(R.id.edit_one);
        btn_add.setOnClickListener(newOnClickListener(){
            @Override
            public void onClick(View v){
```

```
        SpannableString spanStr=newSpannableString("imge");
        Drawable drawable = MainActivity.this.getResources().getDrawable(R.
            drawable.f045);
        drawable.setBounds(0,0,drawable.getIntrinsicWidth(),drawable.
            getIntrinsicHeight());
        ImageSpan span=newImageSpan(drawable,ImageSpan.ALIGN_BANDROID
            STUDIOELINE);
        spanStr.setSpan(span,0,4,Spannable.SPAN_EXCLUSIVE_EXCLUSIVE);
        int cursor=edit_one.getSelectionStart();
        edit_one.getText().insert(cursor, spanStr);
        }
    });
    }
}
```

3.6.10　带删除按钮的 EditText

在 APP 输入界面上经常看到带删除按钮的编辑框,如图 3-73 所示。

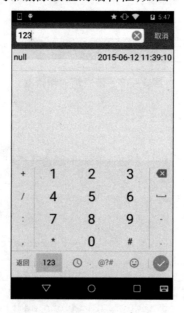

图 3-73　带删除按钮的编辑框

当输入内容后,右面会出现一个"×"图标,点击后会清空输入框中的内容。

开发方式如下:

为 EditText 设置 addTextChangedListener,重写 TextWatcher( )中的抽象方法,用于监听输

入框变化；然后用 setCompoundDrawablesWithIntrinsicBounds 设置"×"的图片；最后重写 onTouchEvent( )方法，如果点击区域是小叉叉图片的位置，清空文本。

实现代码如下：

```java
public clAndroid Studios EditTextWithDel extends EditText {
    private final static String TAG = "EditTextWithDel";
    private Drawable imgInable;
    private Drawable imgAble;
    private Context mContext;
    public EditTextWithDel(Context context) {
        super(context);
        mContext = context;
        init();
    }
    public EditTextWithDel(Context context, AttributeSet attrs) {
        super(context, attrs);
        mContext = context;
        init();
    }
    public EditTextWithDel(Context context, AttributeSet attrs, int defStyleAttr) {
        super(context, attrs, defStyleAttr);
        mContext = context;
        init();
    }
    private void init() {
        imgInable = mContext.getResources().getDrawable(R.drawable.delete_gray);
        addTextChangedListener(new TextWatcher() {
            @Override
            public void onTextChanged(CharSequence s, int start, int before, int
                    count) {
            }
            @Override
            public void beforeTextChanged(CharSequence s, int start, int count, int
                    after) {
            }
            @Override
            public void afterTextChanged(Editable s) {
                setDrawable();
```

```java
            }
        });
        setDrawable();
    }
    // 设置删除图片
    private void setDrawable() {
        if (length() < 1)
            setCompoundDrawablesWithIntrinsicBounds(null, null, null, null);
        else
            setCompoundDrawablesWithIntrinsicBounds(null, null, imgInable, null);
    }
    // 处理删除事件
    @Override
    public boolean onTouchEvent(MotionEvent event) {
        if (imgAble != null && event.getAction() == MotionEvent.ACTION_UP) {
            int eventX = (int) event.getRawX();
            int eventY = (int) event.getRawY();
            Log.e(TAG, "eventX = " + eventX + "; eventY = " + eventY);
            Rect rect = new Rect();
            getGlobalVisibleRect(rect);
            rect.left = rect.right - 100;
            if (rect.contains(eventX, eventY))
                setText("");
        }
        return super.onTouchEvent(event);
    }
    @Override
    protected void finalize() throws Throwable {
        super.finalize();
    }
}
```

3.7 ImageView（图像视图）

ImageView（图像视图）是用来显示图像的控件，本节讲解的内容包括：ImageView 的 src 属性和 blackground 的区别；adjustViewBounds 设置图像缩放时是否按长宽比；scaleType 设置缩放类型;最简单的绘制圆形的 ImageView。

3.7.1 src 属性和 background 属性的区别

在 API 文档中可发现 ImageView 有两个可以设置图片的属性,分别是 src 和 background:

(1) background:通常指的都是背景,而 src 指的是内容。

(2) 当使用 src 填入图片时,是按照图片大小直接填充,并不会进行拉伸;而使用 background 填入图片时,则会根据 ImageView 给定的宽度来进行拉伸。

具体实现代码如下:

```xml
<LinearLayout xmlns:android = "http://schemAndroid Studio.android.com/apk/res/
    android"
  xmlns:tools = "http://schemAndroid Studio.android.com/tools"
  android:id = "@ + id/LinearLayout1"
  android:layout_width = "match_parent"
  android:layout_height = "match_parent"
  android:orientation = "vertical"
  tools:context = "com.jay.example.imageviewdemo.MainActivity" >
<ImageView
  android:layout_width = "wrap_content"
  android:layout_height = "wrap_content"
  android:background = "@drawable/pen" />
<ImageView
  android:layout_width = "200dp"
  android:layout_height = "wrap_content"
  android:background = "@drawable/pen" />
<ImageView
  android:layout_width = "wrap_content"
  android:layout_height = "wrap_content"
  android:src = "@drawable/pen" />
<ImageView
  android:layout_width = "200dp"
  android:layout_height = "wrap_content"
  android:src = "@drawable/pen" />
</LinearLayout>
```

程序运行效果如图 3-74 所示,可以判断程序运行是否正常。

结果分析:宽、高都是 wrap_content,即原图大小,但是当固定了宽或者高时,差别就显而易见,blackground 完全填充了整个 ImageView,而 src 依旧那么大,而且会居中显示,这就涉及 ImageView 的另一个属性 scaleType。这里只设置 width 或者 height,假如同时设置了 width 和 height,blackground 依旧填充,但是 src 的大小可能发生改变,例如测试下下面这段代码:

```
<ImageView
    android:layout_width = "100dp"
    android:layout_height = "50dp"
    android:src = "@drawable/pen"/>
```

程序运行效果如图 3-75 所示,可以判断程序运行是否正常。

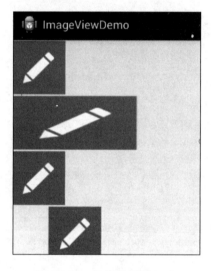

图 3-74 运行效果图(一)

图 3-75 运行效果图(二)

3.7.2 解决 blackground 拉伸导致图片变形的方法

在图 3-74 所示效果图的第二个 Imageview 中可以看到图片已经被拉伸变形,正方形变成了长方形,这种情况并不是我们想要的效果,解决方式有以下两种:

(1)适用于动态加载 ImageView,代码并不复杂,只要在添加 View 时,把大小固定即可:

```
LinearLayout.LayoutParams layoutParam = newLinearLayout.LayoutParams(48,48);
    layout.addView(ibtnPen, layoutParam);
```

(2)除了动态加载 View,更多的时候,还是会通过 xml 布局的方式引入 ImageView 的解决方法,通过 drawable 的 Bitmap 资源文件来实现,然后 blackground 属性设置为该文件即可。这个 xml 文件在 drawable 文件夹下创建,这个文件夹需要手动创建。

pen_bg.xml:
```
<bitmap
    xmlns:android = "http://schemAndroid Studio.android.com/apk/res/android"
    android:id = "@id/pen_bg"
    android:gravity = "top"
    android:src = "@drawable/pen"
    android:tileMode = "disabled" > </bitmap >
```

上述代码并不难理解,titleMode 属性是平铺,即多个小图标铺满整个屏幕。

调用方法如下:

```
ibtnPen.setBacklgroundResource(R.drawable.penbg);        //动态
android:background = "@drawable/penbg"                   //静态
```

3.7.3　src 和 background 结合应用

图 3-76 所示为一个简单的 GridView(导航布局界面),每个项目都是一个 ImageView,但是,上面的图标都不是规则的,如圆形、圆角矩形等,于是这里用到了 src + background。要实现图中效果,需要两个操作:找一张透明的 png 图片 + 设置一个黑色的背景(也可以设置 png 的透明度来实现,但结果可能和预想的有出入)。这里写个简单例子实现呆萌的小猪效果,如图 3-77 所示。

图 3-76　GridView 图

图 3-77　示例 Imageview + background

实现代码如下:

```
<ImageView
    android:layout_width = "150dp"
    android:layout_height = "wrap_content"
    android:src = "@drawable/pig"
    android:background = "#6699FF"/>
```

Java 代码中设置 blackground 和 src 属性：

```
setImageDrawable();          //前景(对应 src 属性)
setBackgroundDrawable();//背景(对应 background 属性)
```

3.7.4 adjustViewBounds 设置缩放是否保存原图长宽比

ImageView 提供了 adjustViewBounds 属性，用于设置缩放时是否保持原图长宽比，单独设置不起作用，需要配合 maxWidth 和 maxHeight 属性一起使用，而后面这两个属性需要 adjustViewBounds 为 true 才会生效。

（1）android:maxHeight：设置 ImageView 的最大高度。

（2）android:maxWidth：设置 ImageView 的最大宽度。

示例代码：

```
<LinearLayout xmlns:android = "http://schemAndroid Studio.android.com/apk/res/
    android"
    xmlns:tools = "http://schemAndroid Studio.android.com/tools"
    android:layout_width = "match_parent"
    android:layout_height = "match_parent"
    android:orientation = "vertical"
    tools:context = ".MainActivity" >
<!-- 正常的图片 -- >
<ImageView
    android:id = "@ + id/imageView1"
    android:layout_width = "wrap_content"
    android:layout_height = "wrap_content"
    android:layout_margin = "5px"
    android:src = "@mipmap/meinv"/>
<!-- 限制了最大宽度与高度,并且设置了调整边界来保持所显示图像的长宽比 -- >
```

```
<ImageView
    android:id = "@ + id/imageView2"
    android:layout_width = "wrap_content"
    android:layout_height = "wrap_content"
    android:layout_margin = "5px"
    android:adjustViewBounds = "true"
    android:maxHeight = "200px"
    android:maxWidth = "200px"
    android:src = "@mipmap/meinv"/ >
</LinearLayout >
```

程序运行效果如图 3 - 78 所示。

图 3 - 78　程序运行效果

结果分析:大的图片是没有进行任何处理的图片,尺寸是 541 × 374 像素;小的图片通过 maxWidth 和 maxHeight 限制 ImageView 最大宽度与高度为 200 px。

3.7.5　scaleType 设置缩放类型

android:scaleType 用于设置显示的图片如何缩放或者移动以适应 ImageView 的大小。Java 代码中可以通过 imageView. setScaleType(ImageView. ScaleType. CENTER)来设置,可选值如下:

(1)fitXY:对图像的横向与纵向进行独立缩放,使得该图片完全适应 ImageView,但是图片的横纵比可能会发生改变。

(2)fitStart:保持纵横比缩放图片,只到较长的边与 Image 的边相等,缩放完成后将图片放在 ImageView 的左上角。

(3) fitCenter：同上，缩放后放于中间。

(4) fitEnd：同上，缩放后放于右下角。

(5) center：保持原图的大小，显示在 ImageView 的中心。当原图的尺寸大于 ImageView 的尺寸时，超过部分进行裁剪处理。

(6) centerCrop：保持横纵比缩放图片，只到完全覆盖 ImageView，可能会出现图片显示不完全的情况。

(7) centerInside：保持横纵比缩放图片，直到 ImageView 能够完全显示图片。

(8) matrix：默认值，不改变原图的大小，从 ImageView 的左上角开始绘制原图，原图超过 ImageView 的部分做裁剪处理。

接下来进行一组对比：

1. 1 fitEnd、fitStart、fitCenter

这里以 fitEnd 为例，其他两个类似。

示例代码：

```
<!-- 保持图片的横纵比缩放,该图片能够显示在ImageView组件上,并将缩放好的图片显示在
    imageView 的右下角 -->
<ImageView
    android:id = "@ + id/imageView3"
    android:layout_width = "300px"
    android:layout_height = "300px"
    android:layout_margin = "5px"
    android:scaleType = "fitEnd"
    android:src = "@mipmap/meinv"/>
```

程序运行效果如图 3-79 所示。

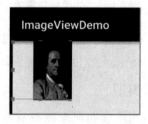

图 3-79　程序运行效果图

2. centerCrop 与 centerInside

(1) centerCrop：按横纵比缩放，直接完全覆盖整个 ImageView。

(2) centerInside：按横纵比缩放，使得 ImageView 能够完全显示这张图片。

实现代码如下:

```xml
<ImageView
    android:layout_width="300px"
    android:layout_height="300px"
    android:layout_margin="5px"
    android:scaleType="centerCrop"
    android:src="@mipmap/meinv"/>

<ImageView
    android:layout_width="300px"
    android:layout_height="300px"
    android:layout_margin="5px"
    android:scaleType="centerInside"
    android:src="@mipmap/meinv"/>
```

程序运行效果如图3-80所示。

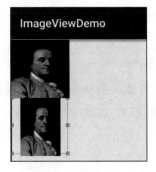

图3-80 程序运行效果

3. fitXY

不按比例缩放图片,目标是把图片塞满整个视图。

示例代码:

```xml
<ImageView
    android:layout_width="300px"
    android:layout_height="300px"
    android:layout_margin="5px"
    android:scaleType="centerCrop"
    android:src="@mipmap/meinv"/>
```

程序运行效果如图 3-81 所示。

图 3-81　程序运行效果

小　　结

本章介绍了 Java 环境 JDK 的搭建和 Android Studio 软件的安装、如何用 Android Studio 创建新的项目、程序的签名打包，以及各种基本组件的应用。

习　　题

1. 编写一个 Android 程序，包含登录注册功能，可以在手机上正常运行。
2. 尝试一下给 Android 程序不同签名，观察安装过程中会遇到什么问题。

第 4 章

可穿戴实验平台蓝牙通信设计

本章通过运用实例学习 Android 蓝牙方面的知识、相关界面的制作和代码的编写,以及如何通过蓝牙实现 Android 设备之间的连接和通信。

学习目标

- 能够使用 Android 创建相关界面和组件。
- 掌握 Android 蓝牙开发方面的技术。

4.1 Android 菜单界面创建与 Fragment 使用

4.1.1 Fragment 的创建

下面介绍一下主界面 MainFragment 的创建:

如图 4 – 1 所示,单击 app 项目,自动新建 Fragment 包,用于存放所有的 Fragment 文件。

(1)新建的 fragment_main.xml,作为项目的主界面布局 UI。即 MainFragment 的布局 UI,实现代码如下:

```
<? xml version = "1.0" encoding = "utf - 8"? >
<LinearLayout xmlns:android = "http://schemAndroid
    Studio.android.com/apk/res/android"
    android:layout_width = "match_parent"
    android:orientation = "vertical"
    android:layout_height = "match_parent" >
```

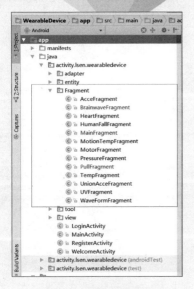

图 4 – 1　文件列表图

```
<GridView                        //定义布局
    android:paddingTop = "10dp"
    android:id = "@ + id/main_gridview"
    android:layout_width = "match_parent"
    android:layout_height = "0dp"
    android:layout_weight = "9"
    android:columnWidth = "80dp"
    android:gravity = "center"
    android:listSelector = "#00000000"
    android:numColumns = "3"
    android:scrollbars = "none"
    android:stretchMode = "columnWidth"
    android:verticalSpacing = "10dp" />
<TextView                        //定义一个文本框
    android:id = "@ + id/main_stateTv"
    android:layout_width = "match_parent"
    android:layout_height = "0dp"
    android:layout_marginTop = "10dp"
    android:layout_weight = "1"
    android:gravity = "center"
    android:textColor = "@color/text_color"
    android:textSize = "@dimen/text_title" />

</LinearLayout>
```

fragment_main. xml 运行效果如图 4 – 2 所示。

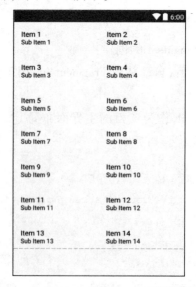

图 4 – 2 fragment_main. xml 运行效果

（2）新建 MainFragment 继承 Fragment。实现代码如下：

```
public clAndroid Studios MainFragment extends Fragment {
    @Override
    public void onAttach(Context context) {                //程序初始化入口
        super.onAttach(context);
        this.context = context;
        this.mainActivity = (MainActivity) context;
    }

    @Override                                              //界面调用函数
    public void onCreate(@Nullable Bundle savedInstanceState) {
        super.onCreate(savedInstanceState);
        FmainHandler = new FmainHandler(mainActivity.mainFragment);
    }
}
```

4.1.2　GridView 的使用

（1）初始化 GridView，并通过 GridView 的 ItemClickListener( ) 得到当前点击的是第几个 Item。

实现代码如下：

```
private void initGridView() {                              //初始化定义的GridView
    data_list = new ArrayList< >();
    initItemData();
    GridViewAdapter gridViewAdapter = new GridViewAdapter(data_list, context);
    gridview.setAdapter(gridViewAdapter);
    gridview.setOnItemClickListener(new AdapterView.OnItemClickListener() {
        @Override
        public void onItemClick(AdapterView<?> parent, View view, int position,
                long id) {
            mainActivity.onClickedFragment(position);      //设置GridView的位置
        }
    });                                                    //设置文本
    stateTv.setText(getResources().getString(R.string.un_connect));
}
```

（2）根据点击的 Item 判断当前所切换的 Fragment。

（3）先判断当前切换的 Fragment 是否已经存在，如果不存在，则实例化一个 Fragment；如果已经存在，则直接显示。实现代码如下：

```java
public void onClickedFragment(int id) {               //设置按钮回调函数
    switch (id) {
        cAndroid Studioe 0:
            if (null == heartFragment) {
                heartFragment = new HeartFragment();
            }
            newFragment = heartFragment;
            titleTv.setText("心率计数实验");              //设置标签显示的文本
            break;
        cAndroid Studioe 1:
            if (null == motorFragment) {
                motorFragment = new MotorFragment();
            }
            newFragment = motorFragment;
            titleTv.setText("腕部触感提醒实验");           //设置标签显示的文本
            break;
        cAndroid Studioe 2:
            if (null == acceFragment) {
                acceFragment = new AcceFragment();
            }
            newFragment = acceFragment;
            titleTv.setText("手掌运动轨迹记录实验");        //设置标签显示的文本
            break;
        cAndroid Studioe 3:
            if (null == tempFragment) {
                tempFragment = new TempFragment();
            }
            newFragment = tempFragment;
            titleTv.setText("体温检测实验");              //设置标签显示的文本
            break;
        cAndroid Studioe 4:
            if (null == brainwaveFragment) {
                brainwaveFragment = new BrainwaveFragment();
            }
            newFragment = brainwaveFragment;
            titleTv.setText("脑电波检测实验");             //设置标签显示的文本
```

```
        break;
}
```

(4) 自定义 adapter——GridViewAdapter：作为 GridView 的适配器，实现代码如下：

```
public clAndroid Studios GridViewAdapter extends BAndroid StudioeAdapter {
    private List<Map<String, Object>> data_list;
    private Context context;
    public GridViewAdapter (List<Map<String, Object>> data_list, Context context) {
        this.data_list = data_list;          //函数赋值
        this.context = context;
    }
    }
    static clAndroid Studios ViewHolder {     //设置 view 中具体的文本信息
        @BindView(R.id.grid_item_imageview)
        ImageView gridItemImageview;
        @BindView(R.id.grid_item_textview)
        TextView gridItemTextview;
        ViewHolder(View view) {
            ButterKnife.bind(this, view);
        }
    }
```

(5) 自定义 GridViewAdapter 的布局 UI——fragment_gridview，实现代码如下：

```
<?xml version="1.0" encoding="utf-8"?>
<LinearLayout xmlns:android="http://schemAndroid Studio.android.com/apk/res/
    android"                      //初始化界面基本设置
    android:layout_width="wrap_content"
    android:layout_height="wrap_content"
    android:gravity="center"
    android:orientation="vertical"
    android:padding="10dp">
<LinearLayout                     //设置 Layout
    android:layout_width="@dimen/gridview_height"
    android:layout_height="@dimen/gridview_height"
    android:background="@drawable/shape_gridview">
```

```
<ImageView            //设置 ImageView
    android:contentDescription = "@null"
    android:id = "@+id/grid_item_imageview"
    android:layout_width = "@dimen/gridview_height"
    android:layout_height = "@dimen/gridview_height"
    android:background = "@mipmap/item_1"
    />
</LinearLayout>
<TextView             //设置 TextView
    android:id = "@+id/grid_item_textview"
    android:layout_width = "match_parent"
    android:layout_height = "wrap_content"
    android:layout_marginTop = "5dp"
    android:gravity = "center"
    android:text = "@string/item_heart"
    android:textColor = "@color/text_color" />
</LinearLayout>
```

子项心率界面如图 4-3 所示。

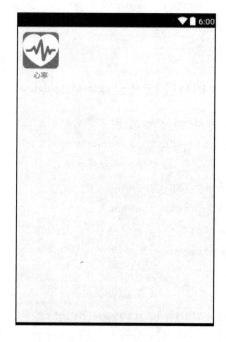

图 4-3　心率界面

(6) 在 MainActivity 中创建 MainFragment,实现代码如下:

```
private void init() {
    Intent intent = getIntent();
    user = (User) intent.getSerializableExtra("User");
    mainHandler = new MainHandler(this);
    if (currentFragment == null) {          //判断变量是否为空值
        mainFragment = new MainFragment();
        currentFragment = mainFragment;
        getSupportFragmentManager().beginTransaction().replace(R.id.main_fragment,
            currentFragment).commit();
        frontFragment = currentFragment;
    }
}
```

MainFragment 主界面如图 4-4 所示。

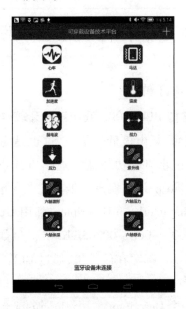

图 4-4　MainFragment 主界面

4.1.3　Fragment 的切换

创建多个 Fragment,当在 MainFragment 中点击不同的可穿戴模块时,切换到不同的模块界面。当前需要切换 Fragment 时,如果没有创建则创建;如果已经创建则隐藏当前 Fragment,显示切换的 Fragment。

Fragment 的切换代码如下:

```java
public void switchFragment(Fragment front, Fragment current) {
    FragmentTransaction fragmentTransaction = getSupportFragmentManager().
    beginTransaction();
    if (currentFragment!=current) {
        currentFragment = current;
        if (!current.isAdded()) {           //判断是否点击
            fragmentTransaction.hide(front).add(R.id.main_fragment, current).
            commit();
        } else {
            fragmentTransaction.hide(front).show(current).commit();
        }
        frontFragment = currentFragment; // 显示切换的 Fragment
    }
}
```

4.2 Android 蓝牙 4.0 通信与通信协议设计

4.2.1 Android 蓝牙 4.0 详解

Android 4.3 及以上版本才支持 BLE API。低功耗蓝牙较传统蓝牙,具有传输速度更快、覆盖范围更广、安全性更高、延迟更短、耗电极低等优点,这也是为什么近年来智能穿戴的东西越来越多、越来越火的原因。传统蓝牙与低功耗蓝牙通信方式也有所不同,传统蓝牙一般通过 Socket 方式,而低功耗蓝牙通过 Gatt 协议来实现。低功耗蓝牙也称 BLE,下面都称为 BLE。

BLE 分为三部分:Service、Characteristic、Descriptor,都用 UUID 作为唯一标识符。UUID 的这种格式为:0000ffe1-0000-1000-8000-00805f9b34fb。例如,有 3 个 Service,就有 3 个不同的 UUID 与 Service 对应。这些 UUID 都写在硬件里,通过 BLE 提供的 API 可以读取。

一个 BLE 终端可以包含多个 Service,一个 Service 可以包含多个 Characteristic,一个 Characteristic 包含一个 value 和多个 Descriptor,一个 Descriptor 包含一个 Value。Characteristic 是比较重要的,是手机与 BLE 终端交换数据的关键,读取设置数据等操作都涉及 Characteristic 的相关属性。

4.2.2 蓝牙 4.0 开发

蓝牙扫描实现的基本步骤如下:
(1)添加权限。
(2)判断设备是否支持 BLE。

(3) 获取蓝牙适配器 BluetoothAdapter。
(4) 判断蓝牙是否开启,如果未开启,则弹出开启窗口。
(5) 开启后,开始进行蓝牙扫描。
(6) 通过 BluetoothAdapter.LeScanCallback 获取扫描结果。

1. 开启蓝牙权限

在 AndroidManifest.xml 中添加蓝牙权限:

```
<uses-permission android:name="android.permission.BLUETOOTH"/>
<uses-permission android:name="android.permission.BLUETOOTH_ADMIN"/>
```

如果想使应用程序只能在支持 BLE 的设备上运行,可以将下面的声明包含进应用程序 manifest 文件中:

```
<uses-feature android:name="android.hardware.bluetooth_le" android:required="true"/>
```

但是,如果想让应用程序也能够在不支持 BLE 的设备上运行,就应该将上面标签中的属性设置为 required="false",然后在运行过程中使用 PackageManager.hAndroid StudioSystemFeature()方法来判断设备是否支持 BLE。

2. 获取蓝牙适配器 BluetoothAdapter

实现代码如下:

```
private boolean isBleEnabled() {
    BluetoothManager bluetoothManager = (BluetoothManager) getSystemService(Context.BLUETOOTH_SERVICE);
    bluetoothAdapter = bluetoothManager.getAdapter();
    return bluetoothAdapter != null && bluetoothAdapter.isEnabled();
}
```

3. 开启蓝牙设备

实现代码如下:

```
private void enableBluetooth(){
    if (bluetoothAdapter == null || !bluetoothAdapter.isEnabled()){
        Intent intent = new Intent(BluetoothAdapter.ACTION_REQUEST_ENABLE);
        startActivityForResult(intent, 2);
    }
}
```

4. 蓝牙工作原理

开始扫描蓝牙设备,如果传入的参数是 true,则扫描指定的 UUID,反之则扫描所有的蓝牙设备。蓝牙设备不能无限期地扫描,扫描一定时间后(10 s)要停止扫描。避免设备不在可用范围时持续不停地扫描,消耗电量。可穿戴设备 UUID:

```
private final static UUID SERVICE_UUID = UUID
    .fromString("6e400001-b5a3-f393-e0a9-e50e24dcca9e");
private final static UUID TX_SERVICE_UUID = UUID
    .fromString("6e400003-b5a3-f393-e0a9-e50e24dcca9e");
private final static UUID RX_SERVICE_UUID = UUID
    .fromString("6e400002-b5a3-f393-e0a9-e50e24dcca9e");
private static final UUID CLIENT_CHARACTERISTIC_CONFIG_DESCRIPTOR_UUID = UUID
    .fromString("00002902-0000-1000-8000-00805f9b34fb");
```

(1)点击扫描:扫描到设备后则在 Dialog 上显示。当点击其中的某一个蓝牙设备后,需要停止当前的扫描,减少手机的能耗。

(2)开始扫描:项目中扫描的是所有的蓝牙设备。扫描 10 s 后自动停止扫描,若再次扫描,则需要点击 Dialog 对话框中的扫描按钮。实现代码如下:

```
private void scannerDialog(){
    bluetoothDevice = ((ExtendedBluetoothDevice) deviceListAdapter
        .getItem(position)).device;
    connectGatt(bluetoothDevice);    //选择蓝牙设备
    dialog.cancel();
    }
}
mScanButton.setOnClickListener(new View.OnClickListener(){
    @Override
    public void onClick(View v){
        if (v.getId() == R.id.action_cancel){
            mainHandler.sendEmptyMessage(7);
            if (isScanning) {        //设置弹出对话框
                if (dialog!=null)
                    dialog.cancel();
            } else {
                startScan(false);
            }
```

```
        }
    }
});
dialog.show();
startScan(false);                    //开启蓝牙
}
private void startScan(boolean is) {
    menuLayout.setEnabled(false);
    if (is) {
        UUID[] uuids = new UUID[1];
        uuids[0] = SERVICE_UUID;
        bluetoothAdapter.startLeScan(uuids, scanCallback);
    } else {
        bluetoothAdapter.startLeScan(scanCallback);
    }
    mScanButton.setText(getResources().getString(R.string.cancel));
    isScanning = true;
    mainHandler.postDelayed(new Runnable() {
        @Override
        public void run() {
            if (isScanning) {
                stopScan();              //关闭蓝牙
            }
        }
    }, dem);
}
```

(3)停止扫描。实现代码如下：

```
private void stopScan() {
    if (isScanning) {                    //判断是否需要扫描
        bluetoothAdapter.stopLeScan(scanCallback);
        isScanning = false;
        showToAndroid Studiot(getResources().getString(R.string.end_scan));
    }
    mScanButton.setText(getResources().getString(R.string.scan));
    menuLayout.setEnabled(true);      //设置蓝牙开启
}
```

5. 回调 LeScanCallback 接口

搜索到设备会回调 LeScanCallback 接口,扫描到蓝牙设备后保存起来并显示。实现代码如下:

```
private BluetoothAdapter.LeScanCallback scanCallback = new BluetoothAdapter.LeScanCallback() {
    @Override
    public void onLeScan(BluetoothDevice device, int rssi, byte[] scanRecord) {
        if (device! = null) {
            updateScannedDevice(device, rssi);
            addScannedDevice(device, rssi, DEVICE_NOT_BONDED);
        }
    }
};
```

6. 连接指定的蓝牙设备

连接后会返回一个 BluetoothGatt 类型的对象,这里定义为 bluetoothGatt。该对象比较重要,后面发现服务读写设备等操作都是通过该对象。回调有 10 个方法,用来处理蓝牙的各种状态,可以根据实际情况实现其中部分方法。实现代码如下:

```
private void connectGatt(BluetoothDevice bluetoothDevice){
    if (bluetoothGatt! = null) {
        closeBluetoothGatt();          //判断蓝牙状态
    }
    if (bluetoothGatt == null) {
        bluetoothGatt =bluetoothDevice.connectGatt(MainActivity.this, false, bluetoothGattCallback);          //处理蓝牙状态
    }
}
```

7. 回调 BluetoothGattCallback 接口中的函数

(1)设备连接成功并回调 BluetoothGattCallback 接口中的 onConnectionStateChange( )函数,然后调用 bluetoothGatt. discoverServices( );去发现服务。发现服务后会回调 BluetoothGattCallback 接口中的 onServicesDiscovered( )函数。

```java
private BluetoothGattCallback bluetoothGattCallback = new BluetoothGattCallback() {
    @Override
    public void onConnectionStateChange (BluetoothGatt gatt, int status, int
            newState) {
        super.onConnectionStateChange(gatt, status, newState);
        if (status == BluetoothGatt.GATT_SUCCESS) {
            if (newState == BluetoothProfile.STATE_CONNECTED) {
                bluetoothGatt.discoverServices();
            } else if (newState == BluetoothProfile.STATE_DISCONNECTED) {
                disconnect();                    //关闭连接
                closeBluetoothGatt();            //关闭蓝牙
                if (!isOut) {
                    mainFragment.FmainHandler.sendEmptyMessage(2);    //发送信息
                }
                isConnectState = false;
                mainHandler.sendEmptyMessage(6);
            }
        } else {
            disconnect();                        //关闭连接
            closeBluetoothGatt();                //关闭蓝牙
            mainFragment.FmainHandler.sendEmptyMessage(2);
            isConnectState = false;
            mainHandler.sendEmptyMessage(6);
        }
    }
```

（2）在onServicesDiscovered( )函数获得了Characteristic后，调用setCharacteristicNotification( )函数，启用该Characteristic的通知。实现代码如下：

```java
public void onServicesDiscovered(BluetoothGatt gatt, int status) {
    super.onServicesDiscovered(gatt, status);
    boolean isUart = false;
    if (status == BluetoothGatt.GATT_SUCCESS) {    //判断设备是否有蓝牙权限
        List<BluetoothGattService> services = gatt.getServices();
        for (BluetoothGattService service : services) {
            isUart = true;
```

```
            txCharacteristic = service.getCharacteristic(TX_SERVICE_UUID);
            rxCharacteristic = service.getCharacteristic(RX_SERVICE_UUID);
        }
    if (isUart){                                                    //判断
        boolean is = setCharacteristicNotification(txCharacteristic, true);
    if (is) {
            mainHandler.sendEmptyMessage(5);
            connectDevice = bluetoothDevice;
            isConnectState = true;
             mainFragment.FmainHandler.sendEmptyMessage(1);    //发送空消息
            }
        }
            }
    }
```

(3) 启用 txCharacteristic 的通知,获得蓝牙设备数据自动返回。实现代码如下:

```
private boolean setCharacteristicNotification(BluetoothGattCharacteristic
characteristic, boolean enable) {
    if (bluetoothGatt == null) {
        return false;
    }          //获得蓝牙设备数据自动返回
    bluetoothGatt.setCharacteristicNotification(characteristic, enable);
}
```

4.2.3 可穿戴蓝牙的数据通信

1. 蓝牙返回数据

蓝牙 BLE 自动返回数据,在 onCharacteristicChanged( ) 函数中回调。

```
@Override
public void onCharacteristicChanged(BluetoothGatt gatt, BluetoothGattCharacteristic
characteristic) {
    super.onCharacteristicChanged(gatt, characteristic);
    if (characteristic.getUuid().equals(TX_SERVICE_UUID)) {
        value = new byte[20];
        for (int i = 0; i < value.length; i ++) {      //遍历数组
```

```
            value[i] = characteristic.getValue()[i];
        if (!isSend) {
            writeCharacteristic(sendData);
            isSend = true;
        }
        mainHandler.sendEmptyMessage(4);          //获得蓝牙设备 BLe 数据自动返回
        }
    }
};
```

2. 蓝牙发送数据

收发分两个 Characteristic 进行，可穿戴蓝牙通过 txCharacteristic 返回数据；手机蓝牙通过 rxCharacteristic 发送数据。实现代码如下：

```
public boolean writeCharacteristic(byte[] value) {
    boolean isSend = false;
    if (bluetoothGatt! = null && rxCharacteristic! = null) {
        rxCharacteristic.setValue(value);
        isSend = bluetoothGatt.writeCharacteristic(rxCharacteristic);
    }
    return isSend;                              //如果发送数据,返回
}
```

3. 断开连接

当通信结束时，需要在 APP 中主动断开连接，实现代码如下：

```
public void disconnect() {
    if (bluetoothGatt == null) {
        return;
    }
    bluetoothGatt.disconnect();                 //主动断开连接
}
```

4. 蓝牙通信协议

（1）通信方式：蓝牙通信，是一种无线技术标准，可实现固定设备、移动设备和楼宇个人之间的短距离数据交换（使用 2.4~2.485 GHz 的 ISM 波段的 UHF 无线电波）。

（2）协议格式（见表 4-1）：

表 4-1 协议格式

格 式	设 备 号	有效数据长度	有效数据区	填补数据区
长度	1 BYTE	1 BYTE	n BYTE	n BYTE

- 设备号及对应设备（见表 4-2）：

表 4-2 设备号及对应设备

设 备 号	传感器及设备
0x01	心率计
0x02	马达
0x03	加速度计
0x04	温度计
0x05	脑电波计
0x06	拉力计
0x07	压力计
0x08	紫外线计

- 有效数据长度：指有效数据区内数据的长度。
- 有效数据区：放置有效的通信数据。
- 填补数据区：因为每次通信数据包的长度都是固定为 20B，所以当有数据的长度不足时，此区域填充 0 以补全 20B。

（3）具体命令定义：

- 心率计（见表 4-3）

表 4-3 心率计

设备号	长　　度	有效数据区
0x01	0x01	心率 0~140：已测的心率； <0：心率计故障

可穿戴实验平台蓝牙通信设计 第4章

- 马达(见表4-4)

表4-4 马达

设备号	长度	有效数据区	
		振动方式	振动参数
0x02	0x02	0:连续振动 1:短促振动	(1)连续振动:0~255,振动单位取值(每个振动单位约10 ms); (2)短促振动:0~20,短促振动的次数

- 加速度计(见表4-5)

表4-5 加速度计

设备号	有效数据长度	有效数据区
		当前计步值
0x03	0x04	0x0000 0000 ~ 0x7FFF FFFF:让步累加值(从上电开始计)

- 温度计(见表4-6)

表4-6 温度计

设备号	有效数据长度	有效数据区	
		温度整数部分	温度小数部分
0x04	0x02	0~100	0~99

- 脑电波(见表4-7、表4-8)

表4-7 脑电波——A类数据包

设备号	长度	有效数据区					
		类型	Delta	Theta	LowAlpha	HighAlpha	放松度
0x05	0x0E	0x00	3 B	3 B	3 B	3 B	0~100

表4-8 脑电波——B类数据包

设备号	长度	有效数据区					
		类型	LowBate	HighBate	LowGamma	MiddleGamma	专注度
0x05	0x0E	0x01	3 B	3 B	3 B	3 B	0~100

- 拉力计(见表4-9)

表4-9 拉力计

设备号	有效数据长度	有效数据区	
		拉力整数部分	拉力小数部分
0x06	0x02	-20~20	0~255

● 压力计(见表 4-10)

表 4-10 压力计

设 备 号	有效数据长度	有效数据区	
		压力整数部分	压力小数部分
0x07	0x02	-100~100	0~255

● 紫外线(见表 4-11)

表 4-11 紫外线

设 备 号	有效数据长度	有效数据区
		紫外线辐射强度
0x08	0x01	0~100

4.3　Android 蓝牙开发——搜索蓝牙设备

本节和 4.4 节主要讲解搜索蓝牙设备、蓝牙设备之间的连接和蓝牙设备之间的通信 3 个主要模块。掌握了这些知识,就能进行简单的蓝牙开发。

搜索界面比较简单,主要有 3 个按钮,两个文本和一个列表,如图 4-5 所示。三个按钮主要对应打开蓝牙、搜索设备和发送信息 3 个功能;两个文本主要用于提示连接状态和发送的消息以及获取的消息;列表主要展示搜索到的蓝牙设备。

图 4-5　搜索界面图

要用到蓝牙功能,需要先在 manifest 中声明一下蓝牙的使用权限。实现代码如下:

```
<uses-permission android:name="android.permission.BLUETOOTH"/>
<uses-permission android:name="android.permission.BLUETOOTH_ADMIN"/>
<uses-permission android:name="android.permission.ACCESS_COARSE_LOCATION"/>
```

4.3.1 打开蓝牙设备

可以通过 BluetoothAdapter.ACTION_REQUEST_ENABLE 提示用户开启蓝牙,也可以 bTAdatper.enable( )直接开启蓝牙。

实现代码如下:

```
BluetoothAdapter bTAdatper = BluetoothAdapter.getDefaultAdapter();
if(bTAdatper == null){
    ToAndroid Studiot.makeText(this,"当前设备不支持蓝牙功能",ToAndroid Studiot.
        LENGTH_SHORT).show();
}
    if(!bTAdatper.isEnabled()){
      /* Intent i = new Intent(BluetoothAdapter.ACTION_REQUEST_ENABLE);
        startActivity(i);*/
        bTAdatper.enable();
    }
    //开启被其他蓝牙设备发现的功能
    if (bTAdatper.getScanMode()!=BluetoothAdapter.SCAN_MODE_CONNECTABLE_
        DISCOVERABLE) {
            Intent i = new Intent(BluetoothAdapter.ACTION_REQUEST_DISCOVERABLE);
            //设置为一直开启
            i.putExtra(BluetoothAdapter.EXTRA_DISCOVERABLE_DURATION, 0);
            startActivity(i);
        }
```

开启蓝牙后,需要设置蓝牙为可发现状态,让其他设备能够搜索到。

```
putExtra(BluetoothAdapter.EXTRA_DISCOVERABLE_DURATION, 0);
```

最后的参数设置为 0,可以让蓝牙设备一直处于可发现状态。当需要设置具体可被发现的时间时,最多只能设置 300s。

```
i.putExtra(BluetoothAdapter.EXTRA_DISCOVERABLE_DURATION, 300);
```

4.3.2 搜索蓝牙设备

成功开启蓝牙设备后,调用蓝牙适配器的 startDiscovery()方法就可以搜索附近可连接的蓝牙设备。另外,可以调用 cancelDiscovery()来取消搜索,实现代码如下:

```
mBluetoothAdapter.startDiscovery();
mBluetoothAdapter.cancelDiscovery()
```

当开始搜索附近蓝牙设备时,系统会发出 3 个搜索状态的广播:BluetoothAdapter. ACTION_DISCOVERY_STARTED、BluetoothDevice. ACTION_FOUND、BluetoothAdapter. ACTION_DISCOVERY_FINISHED。

这 3 项分别对应开始搜索、搜索到设备、搜索结束,因此,可以定义一个广播来获取这些状态。实现代码如下:

```
private final BroadcAndroid StudiotReceiver mReceiver = new BroadcAndroid
    StudiotReceiver() {
  @Override
  public void onReceive(Context context, Intent intent) {
    String action = intent.getAction();
    if (BluetoothDevice.ACTION_FOUND.equals(action)) {
      BluetoothDevice device = intent.getParcelableExtra(BluetoothDevice.
        EXTRA_DEVICE);
      //避免重复添加已经绑定过的设备
      if (device.getBondState()!=BluetoothDevice.BOND_BONDED) {
        //此处的 adapter 是列表的 adapter,不是 BluetoothAdapter
        adapter.add(device);
        adapter.notifyDatAndroid StudioetChanged();
      }
    } else if (BluetoothAdapter.ACTION_DISCOVERY_STARTED.equals(action)) {
      ToAndroid Studiot.makeText(MainActivity.this,"开始搜索",ToAndroid
        Studiot.LENGTH_SHORT).show();
    } else if (BluetoothAdapter.ACTION_DISCOVERY_FINISHED.equals(action)) {
      ToAndroid Studiot.makeText(MainActivity.this,"搜索完毕",ToAndroid
        Studiot.LENGTH_SHORT).show();
    }
  }
};
```

在此,将搜索到的设备添加到列表中进行展示。

```
if (device.getBondState()!=BluetoothDevice.BOND_BONDED)
```

系统会保存之前配对过的蓝牙设备,这里对搜索到的设备进行过滤,判断设备是否已经配对过。因此,还可以直接获取之前配对过的设备。

4.3.3 获取配对过的蓝牙设备

实现代码如下:

```
Set<BluetoothDevice> pairedDevices=bTAdatper.getBondedDevices();
    if (pairedDevices.size()>0) {
        for (BluetoothDevice device : pairedDevices) {
            adapter.add(device);
        }
    }
```

至此,就成功地获取到附近的蓝牙设备。

4.4 Android 蓝牙开发——连接蓝牙设备

上一节中将搜索到的蓝牙设备展示在列表中,现在为列表项增加点击事件。获取到设备后就可以开始处理蓝牙设备之间的连接,实现代码如下:

```
listView.setOnItemClickListener(new AdapterView.OnItemClickListener() {
    @Override
    public void onItemClick(AdapterView<?> parent, View view, int position, long id)
    {
        if (bTAdatper.isDiscovering()) {
            bTAdatper.cancelDiscovery();
        }
        BluetoothDevice device=(BluetoothDevice) adapter.getItem(position);
        //连接设备
        connectDevice(device);
    }
});
```

当点击了列表项后,如果蓝牙设备当前处于搜索状态,则取消搜索。获取到列表项相对应的蓝牙设备后调用连接方法。

连接蓝牙设备,实现代码如下:

```java
private void connectDevice(BluetoothDevice device) {
    text_state.setText(getResources().getString(R.string.connecting));
    try {
        //创建 Socket
        BluetoothSocket socket = device.createRfcommSocketToServiceRecord(BT_UUID);
        //启动连接线程
        connectThread = new ConnectThread(socket, true);
        connectThread.start();
    } catch (IOException e) {
        e.printStackTrace();
    }
}
```

可以看到,在 connectDevice( )方法中,首先用一个 Text 控件提示连接状态,然后用选择的设备调用 createRfcommSocketToServiceRecord(UUID uuid)获取 BluetoothSocket,最后开启一个线程去处理 BluetoothSocket。

这里需要注意的是传入的 UUID。UUID 是 Universally Unique Identifier 的缩写,它是在一定的范围内(从特定的名字空间到全球)唯一的机器生成的标识符。在连接蓝牙设备时,必须确保两个设备的 UUID 是相同的。通常 UUID 是 16B(128 位)长的数字,通常以 36B 的字符串表示:3F2504E0 - 4F89 - 11D3 - 9A0C - 0305E82C3301。

实现代码如下:

```java
//如果是自动连接 则调用连接方法
    if (activeConnect) {
        socket.connect();
    }
    text_state.post(new Runnable() {
        @Override
        public void run() {
//设置显示文本
text_state.setText(getResources().getString(R.string.connect_success));
        }
});
```

在创建线程时会传入布尔变量 activeConnect 做标识,区分是自动连接还是被动连接。如果是自动连接状态,则调用 BluetoothSocket 的 connect( )进行连接。

而被动连接的做法主要是开启一个监听线程,监听是否有设备连接到我们的设备上,实现代码如下:

```
listenerThread = new ListenerThread();
    listenerThread.start();
```

开启监听线程:

```
private BluetoothServerSocket serverSocket;
    private BluetoothSocket socket;
    @Override
    public void run(){
        try {
            serverSocket = bTAdatper.listenUsingRfcommWithServiceRecord(
                NAME, BT_UUID);
            while (true) {
                //线程阻塞,等待别的设备连接
                socket = serverSocket.accept();
                text_state.post(new Runnable() {
                    @Override
                    public void run() {
                            text_state.setText(getResources().getString(R.string.
                                connecting));
                    }
                });
                connectThread = new ConnectThread(socket, false);
                connectThread.start();
            }
        } catch (IOException e) {
            e.printStackTrace();
        }
    }
```

监听线程中的代码很简单,首先是获取 BluetoothServerSocket,这里需要用到 UUID,必须确保和自动连接所用到的 UUID 是相同的,而对于 NAME 就没什么要求。接着就调用 accept( )方法。这个方法会阻塞住当前的线程,直到有设备连接。当有设备连接时,就可以获取到对应的 BluetoothSocket。到此,被动连接就成功了,接着只要把获取到的 BluetoothSocket 交由连接线程 ConnectThread 去处理即可。

小　　结

　　本章通过运用实例讲解了 Android 蓝牙方面的知识、相关界面的制作和代码的编写,以及如何通过蓝牙实现 Android 设备之间的连接和通信,在下一章,将通过几个实际项目来学习可穿戴设备的具体开发技术。

 习　　题

1. 编写一个 Android 蓝牙程序,实现查看已经配对过的蓝牙设备。
2. 蓝牙通信和其他通信方式相比有什么优缺点?

第 5 章

可穿戴设备模块综合设计

本章通过 APP 的开发和介绍,配合可穿戴实验平台模块进行各种实验进行学习,在学习过程中可锻炼学生的动手能力,可以一边进行程序开发,一边进行操作。

学习目标

掌握可穿戴设备项目的设计和开发,并且学习相关 APP 界面和功能的开发。

项目开发平台:

图 5-1 所示为项目中需要用的可穿戴设备的实物图,可以看到硬件平台的几个组成部分。

图 5-1 可穿戴设备实验平台

平台分为各个模块实验区域、显示屏、键盘、开关、数据接口等。

关于APP界面的开发制作,在第3、4章中已经做了介绍,本章就实验平台的实验过程和理论分析进行讲解,并针对APP中各个模块中的功能部分进行讲解和代码分析。

5.1 体温检测模块的开发和设计

5.1.1 体温采集信息价值

1. 人体正常温度

人体体温是生命健康中的一项非常重要的指标,一般孩子的平均体温都在37 ℃左右,成人的平均体温在36.5～36.8 ℃之间;体温的异常表明人体处在一个非正常的状态,或者更严重的疾病状态。体温除了正常状态外,只有高或地两个异常状态,这两种异常状态都是比较危险的。

2. 体温异常对人体的影响

各个年龄段的人群平时的正常体温都具有差别,如图5-2所示。

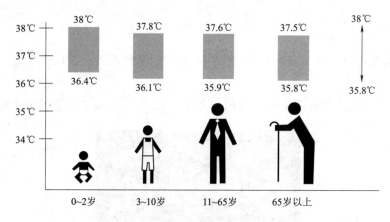

图5-2 正常体温

(1)较高体温。在正常范围内,较高的体温对身体产生如下影响:

● 内脏机能活跃。体温较高,能让体内消化酵素以及对内脏有益的其他酵素更有活力。可预防便秘、胃胀气、尿频等症状。

● 提高免疫力。高体温能提升抵抗外来病毒、细菌的白细胞功能,增强免疫力,改善体质。

● 提高基础代谢率。基础代谢高的人,就算不运动,消耗的热量也比基础代谢低的人要多,较高的体温能提高身体的基础代谢率。

● 血液循环良好。体温高时,血管会变得较为柔软,血液能顺利地输送至全身各处。

(2)较低的体温。在正常范围内,较低的体温对身体产生如下影响:

● 内脏功能低下:体温降低,包括消化酵素在内的有助于内脏活动的各种酵素机能都相应降低;人容易感到疲倦,出现各种不舒服的症状,如便秘、胃胀气、尿量减少等,都会频频出现。

● 免疫力降低:侵入人体的细菌及病毒要由体内的白细胞来对抗,白细胞所提供的免疫力便会减少,因此体温较低的人,在季节交换时比较容易感冒。

● 基础代谢率下降:低体温则使人不易消耗热量,也会让细胞的新陈代谢率衰退,肌肤变差。

● 血液循环变差:低体温的人,手脚的末梢血管会紧缩,血液自然不易流通,更会因心脏输送血液的力量减弱,使全身的血液循环变差。另外,也可能因自律神经发挥的功能降低,导致血管收缩能力受到影响,血液的流通受到阻碍。

5.1.2 非接触式体温传感器原理

1. 非接触式体温测量方法

红外测温传感器是最常用的非接触式测温仪表,基于黑体辐射的基本定律,也被称为辐射测温仪表。一切温度高于绝对零度的物体都在不停地向周围空间发出红外辐射能量。物体的红外辐射能量的大小及其按波长的分布与它的表面温度有着十分密切的关系。因此,通过对物体自身辐射的红外能量的测量,便能准确地测定它的表面温度,这就是红外辐射测温所依据的客观基础。

2. 非接触式体温计构造

红外测温模块由光学系统、光电探测器、信号放大器及信号处理等部分组成。光学系统汇集其视场内的目标红外辐射能量,视场的大小由测温仪的光学零件以及位置决定。红外能量聚焦在光电探测仪上并转变为相应的电信号。该信号经过放大器和信号处理电路按照仪器内部的算法和目标发射率校正后转变为被测目标的温度值。除此之外,还应考虑目标和测温模块所在的环境条件,如温度、气氛、污染和干扰等因素对性能指标的影响及修正方法。非接触式体温计如图 5-3 所示。

图 5-3 非接触式体温计

穿戴设备使用温度传感器型号为 MLX90615,其内部有两颗芯片、红外热电堆探测器和信号处理 Android Studio SP MLX90325,尤其是由 Melexis 设计的处理 IR 传感器输出的芯片。

MLX90325 在信号调节芯片中使用了先进的低噪声放大器,一枚 16 位 ADC(模/数转换器)以及功能强大的 DSP(数字信号处理器)元件,从而实现高精确度温度测量。该传感器计算并存储于 RAM 中的环境温度以及物体温度可实现 0.02°C 的解析度的数据,并且可通过双线标准 IIC 输出获得 (0.02°C 分辨率) 或者通过 10 位 PWM 输出获得。信号调节芯片实物如图 5-4 所示,其原理图如图 5-5 所示。

图 5-4 信号调节芯片

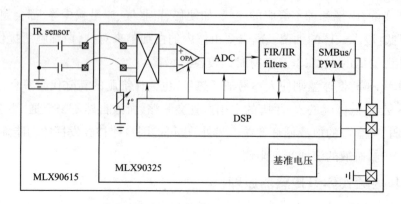

图 5-5　信号调节芯片原理图

5.1.3　温度采集电路解析

MLX90615 有钳位二极管连接在 SDA/SCL 和 VDD 之间。因此需要向 MLX90615 提供电源以使 SMBus 线不成为负载。MLX90615 引脚如图 5-6 所示,它与外围电路连接如图 5-7 所示:

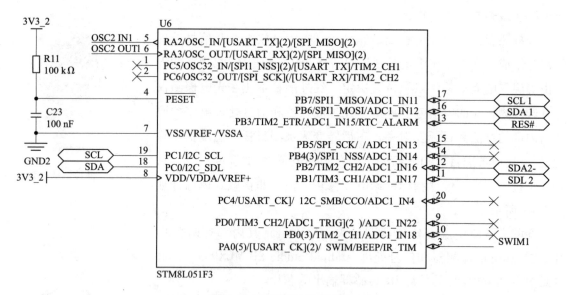

图 5-6　MLX90615 引脚图

5.1.4　体温传感代码解析

1. 体温传感软件功能

MCU 嵌入式系统功能:通过 IIC 总线读出 MLX90615 的数值数据(见图 5-7),并通过 PEC 校验计算得出可靠数据;再从可靠数据中抽取有效的温度数据,并针对温度数据进行转换计算得到摄氏度数据。图 5-8 所示为 MLX90615 电路图。

可穿戴设备模块综合设计 **第 5 章**

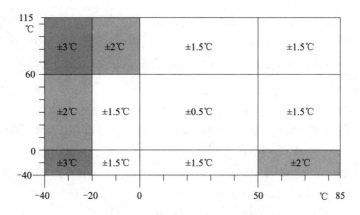

图 5－7　MLX90615 的数值数据

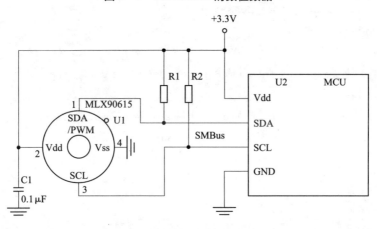

图 5－8　MLX90615 电路图

2. 温度模块的数据解析

温度模块协议体如表 5－1 所示。

表 5－1　温度模块协议体

设　备　号	有效数据长度	有效数据区	
		温度整数部分	温度小数部分
0x04	0x02	0~100	0~99

数据解析：在第 4 章讲到通过 onCharacteristicChanged( ) 函数回调拿到可穿戴模块的数据 20 个 byte 的数组。通过判断 byte 数组的第一个字节可以知道当前返回的模块设备号；判断 byte 数组的第二个字节可以知道当前数组的有效字节长度；判断 byte 数组的第三个字节可以知道温度的整数数值；判断 byte 数组的第四个字节可以知道温度的小数数值。实现代码如下：

```
cAndroid Studioe 0x04:
    if (value[1]==0x02) {
```

```
    int integer=value[2] >=0? value[2] : value[2]+256;
    int decimal=value[3] >=0? value[3] : value[3]+256;
    double dou = Double.valueOf(String.valueOf(integer) + "." + String.valueOf
       (decimal));
    if (activity.weData != null) {
        activity.weData.setTemp(String.valueOf(dou));      //设置数组信息
    }
    if (activity.tempFragment!=null || activity.motionTempFragment != null) {
    if (activity.currentFragment==activity.tempFragment) {
    if (activity.tempFragment.tempHandler != null) {
        Message message = new Message();
        message.what=1;
        message.obj=dou;                                   //赋值
        activity.tempFragment.tempHandler.sendMessage(message);
      }
    } else if (activity.currentFragment==activity.motionTempFragment) {
    activity.motionTempFragment.number=dou;
      }
     }
    }
   break;
```

5.1.5 手机 APP 软件的开发和功能

体温检测:项目 APP 界面的截图如图 5-9 所示。

收集 MCU 发送来的摄氏度数据,并将其存入到数据库中,形成一系列人体体温的历史记录。然后,可以通过调取这些记录描绘出个人历史脉搏曲线。个人历史脉搏曲线如图 5-10 所示。

嵌入式 MLX90615 DSP 控制测量量度,计算物体和环境温度并且进行温度的后处理,将它们通过 IIC 兼容接口或者 PWM 模式输出。

单片机通过 IIC 总线读取 MLX90615 的温度数据,通过 PEC 校验计算得出温度可靠的数据,并将校验后的数据转换为摄氏度数据。其流程图如图 5-11 所示。

项目设计的目的是让大家熟悉体温模块的数据解析,以

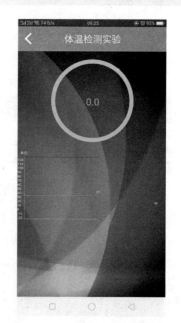

图 5-9 体温检测项目效果图

及使用自定义 RoundProgressBar 视图显示体温数值和自定义 RoundView 视图显示曲线。

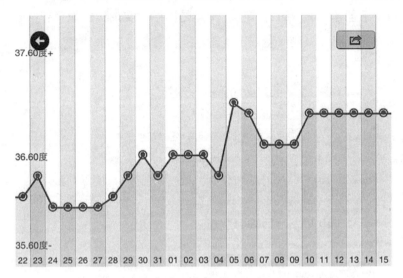

图 5-10　个人历史脉搏曲线

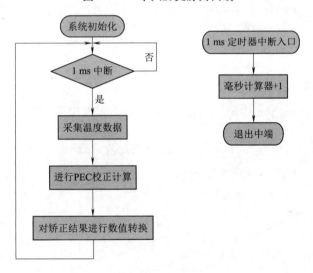

图 5-11　软件流程图

1. 自定义视图

自定义 RoundProgressBar 和自定义 RoundView 显示在 HeartFragment 中。

自定义 RoundProgressBar 的内容包括：

- 自定义 View 的属性。
- 在 View 的构造方法中获得自定义的属性。
- 重写 onMesure（onMesure 不一定是必需的）。
- 重写 onDraw。

(1) 自定义 RoundProgressBar：

● 在 res/values/下创建 attrs.xml；自定义 View 的属性。实现代码如下：

```xml
<?xml version="1.0" encoding="UTF-8"?>
<resources>                                              //自定义view的属性
<declare-styleable name="RoundProgressBar">
<attr name="roundColor" format="color"/>
<attr name="roundProgressColor" format="color"/>
<attr name="roundWidth" format="dimension"/>
<attr name="textColor" format="color"/>                  //设置颜色
<attr name="textSize" format="dimension"/>
<attr name="max" format="integer"/>
<attr name="textIsDisplayable" format="boolean"/>
<attr name="style">
<enum name="STROKE" value="0"/>
<enum name="FILL" value="1"/>
</attr>
</declare-styleable>
</resources>
```

● 在布局 fragment_temp.xml 中声明自定义的 View。实现代码如下：

```xml
<activity.lsen.wearabledevice.tool.RoundProgressBar
    android:layout_marginTop="10dp"
    android:id="@+id/temp_roundbar"                      //设置id
    app:roundWidth="10dp"
    android:layout_width="200dp"
    android:layout_height="200dp"
    android:layout_gravity="center"                      //设置位置
    app:roundColor="@color/round_bar"
    app:textColor="@color/text_color"
    app:textSize="@dimen/text_title"/>
```

● 在 View 的构造方法中获得自定义的属性。实现代码如下：

```java
public RoundProgressBar(Context context){
    this(context, null);
}
public RoundProgressBar(Context context, AttributeSet attrs){
```

```
    this(context, attrs, 0);
}
public RoundProgressBar(Context context, AttributeSet attrs, int defStyle){
    super(context, attrs, defStyle);
    paint=new Paint();            //绘制函数
    TypedArray mTypedArray=context.obtainStyledAttributes(attrs,
    roundWidth=mTypedArray.getDimension(
        R.styleable.RoundProgressBar_roundWidth, 40);
    max=mTypedArray.getInteger(R.styleable.RoundProgressBar_max, 140);
    textIsDisplayable=mTypedArray.getBoolean(
        R.styleable.RoundProgressBar_textIsDisplayable, true);
    style=mTypedArray.getInt(R.styleable.RoundProgressBar_style, 0);
    mTypedArray.recycle();
}
```

- 重写 OnDraw,实现代码如下:

```
@Override
protected void onDraw(CanvAndroid Studio canvAndroid Studio){
    super.onDraw(canvAndroid Studio);
    int centre=getWidth() / 2;
    int radius=(int) (centre-roundWidth / 2);
    paint.setColor(roundColor);
    paint.setStyle(Paint.Style.STROKE);
    paint.setStrokeWidth(roundWidth);
    paint.setAntiAliAndroid Studio(true);
    canvAndroid Studio.drawCircle(centre, centre, radius, paint);
    paint.setStrokeWidth(0);
    paint.setColor(textColor);
    paint.setTextSize(textSize);
    paint.setTypeface(Typeface.DEFAULT_BOLD);
    double percent= (double) (((float) progress / (float) max)*140);
    percent=formatDouble(percent);
    float textWidth=paint.meAndroid StudioureText(percent + company);
    if (textIsDisplayable && percent !=0 && style==STROKE){
        canvAndroid Studio.drawText(percent+company, centre-textWidth / 2, centre
```

```
                + textSize / 2, paint);
        } else {
            canvAndroid Studio.drawText(percent+company, centre-textWidth / 2, centre
                + textSize / 2, paint);
        }
        paint.setStrokeWidth(roundWidth);          //设置线条宽度
        paint.setColor(textColor);
        RectF oval=new RectF(centre-radius, centre-radius, centre
                +radius, centre+radius);
        switch (style){
            cAndroid Studioe STROKE: {
                paint.setStyle(Paint.Style.STROKE);
                canvAndroid Studio.drawArc(oval, 0, (float) (360 * progress / max),
                    false, paint);                 //绘制过程中画线
                break;
            }
            cAndroid Studioe FILL: {
                paint.setStyle(Paint.Style.FILL_AND_STROKE);
                if (progress !=0)
                    canvAndroid Studio.drawArc(oval, 0, (float) (360 * progress / max),
                        true, paint);
                break;
            }
        }
}
```

- RoundProgressBar 效果图如图 5 – 12 所示。

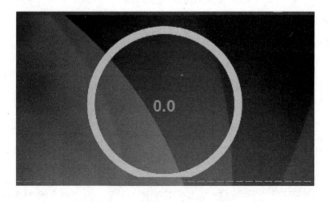

图 5 – 12　RoundProgressBar 效果图

● 在 RoundProgressBar 中设置数值的接口,实现代码如下:

```
public synchronized void setProgress(double progress){
    if (progress>max){
        progress=max;
    }
    if (progress<=max){
        this.progress=progress;
        postInvalidate();
    }
}
```

(2)自定义 RoundView:
该视图未做屏幕适配处理,如果需要可以自己尝试修改。
● 绘制曲线边线的 XY 轴,实现代码如下:

```
@Override
protected void onDraw(CanvAndroid Studio canvAndroid Studio){
    super.onDraw(canvAndroid Studio);
    MaxDatAndroid Studioize=XLength / XScale;
    YLabel=new String[YLength / YScale];
    if (swi==0 || swi==8){
        for (int i=0; i<YLabel.length; i ++){
            YLabel[i]=(i) +"";
        }
    } else {
        for (int i=0; i<YLabel.length; i ++){
            YLabel[i]=(i-20) +"";
        }
    }
    Paint paint=new Paint();
    paint.setStyle(Paint.Style.STROKE);
    paint.setAntiAliAndroid Studio(true);
    paint.setTextSize(25);                    //设置字体大小
    paint.setColor(Color.WHITE);              //设置字体颜色
    Paint paint2=new Paint();
    paint2.setStyle(Paint.Style.STROKE);
    paint2.setAntiAliAndroid Studio(true);
```

```
    paint2.setColor(Color.RED);
    canvAndroid Studio.drawLine(XPoint, YPoint-YLength, XPoint, YPoint, paint);
```

- 根据不同的模块有不同的曲线要求,所以根据模块来绘制。实现代码如下:

```
   if (swi!=8){
   for (int i=0; i * YScale < YLength; i++){
   if (i==0 ||i==10 ||i==20 ||i==30 ||i==40 ||i==50
      ||i==60 ||i==70 ||i==80 ||i==90 ||i==100
      ||i==110 ||i==120 ||i==130 ||i==140){
      canvAndroid Studio.drawLine(XPoint, YPoint-i*YScale, XPoint+5, YPoint
       -i*YScale, paint);
      canvAndroid Studio.drawText(YLabel[i], XPoint-50, YPoint-i*YScale,
      paint);
      } else {
      if(swi==0){
         if(i==75){
            canvAndroid Studio.drawLine(XPoint, YPoint-i*YScale, XPoint+XLength,
            YPoint-i * YScale, paint);          //使用方法,绘制
            canvAndroid Studio.drawText(getResources().getString(R.string.dpm_75),
            XPoint+XLength-20, YPoint-i*YScale-15, paint);
         }
      } else if (swi==4) {
      if (i==57) {
            canvAndroid Studio.drawLine(XPoint, YPoint-i*YScale, XPoint+XLength,
            YPoint-i*YScale, paint);          //使用方法,绘制
            canvAndroid Studio.drawText(getResources().getString(R.string.dpm_37),
            XPoint+XLength-20, YPoint-i*YScale-15, paint);
     }
   }
      }
   }
      } else {
         YScale=4;
         for(int i=0; i*YScale<YLength; i++){
            if(i==0 ||i==10 ||i==20 ||i==30 ||i==40 ||i==50
               ||i==60 ||i==70 ||i==80 ||i==90 ||i==100
         ){
```

```
        canvAndroid Studio.drawLine(XPoint, YPoint-i*YScale, XPoint+5, YPoint
            -i*YScale, paint);
        canvAndroid Studio.drawText(YLabel[i], XPoint-50, YPoint-i*YScale,
            paint);
        }
    }
}
canvAndroid Studio.drawText(getResources().getString(R.string.company)+company
XPoint-30, YPoint-YLength-20, paint);
canvAndroid Studio.drawLine(XPoint, YPoint-YLength, XPoint+XLength, YPoint
    -YLength, paint2);
canvAndroid Studio.drawLine(XPoint, YPoint, XPoint+XLength, YPoint, paint);
```

- 绘制曲线。实现代码如下:

```
if(data.size()>1){
   if(swi==4){
      int xy=60;
      for(int i=1; i<data.size(); i++){          //绘制曲线
      canvAndroid Studio.drawLine((float)(XPoint+(i-1)*XScale),
      (float)(YPoint-data.get(i-1)*YScale)-xy,
      (float)(XPoint+i*XScale),
      (float)(YPoint-data.get(i)*YScale)-xy, paint);
      }
   }else{
      for (int i=1; i<data.size(); i++){          //绘制曲线
         canvAndroid Studio.drawLine(XPoint+(i-1)*XScale,
         YPoint-data.get(i-1)*YScale, XPoint+i*XScale,
         YPoint-data.get(i)*YScale, paint);
      }
   }
}
```

因为曲线需要动态显示,需要不断地更新数据调用 invalidate( )来重写 OnDraw( )函数。实现代码如下:

```
private static clAndroid Studios HearHandler extends Handler {
    private WeakReference<RoundView> mActivityReference;
    //定义曲线
```

```
    HearHandler(RoundView activity){         // 重写 OnDraw
        mActivityReference = new WeakReference < > (activity);
    }
    @Override
    public void handleMessage(Message msg){
        final RoundView activity=mActivityReference.get();
        if (activity != null) {
            if (msg.what==1){
                if (activity.data.size()>= activity.MaxDatAndroid Studioize) {
                    activity.data.remove(0);
                }
                activity.invalidate();
            }
        }
    }
}
```

- 在布局 fragment_hear.xml 中声明自定义的 View。实现代码如下：

```
<activity.lsen.wearabledevice.view.RoundView
    android:id = "@ + id/hear_roundview"
    android:layout_width = "match_parent"
    android:layout_height = "0dp"
    android:layout_gravity = "center"
    android:layout_marginTop = "10dp"
    android:layout_weight = "2" />
```

最终效果如图 5-13 所示。

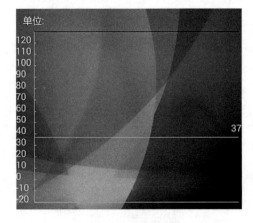

图 5-13 温度曲线图

2. TempFragment 中使用 RoundProgressBar 和 RoundView

（1）在 onCreateView 中使用视图。实现代码如下：

```
@Override
public View onCreateView(LayoutInflater inflater, @Nullable ViewGroup container, @Nullable Bundle savedInstanceState){
    View view=inflater.inflate(R.layout.fragment_temperature, container, false);
    unbinder=ButterKnife.bind(this, view);         //使用视图设置位置
    tempHandler=new TempHandler(mainActivity.tempFragment);
    tempRoundview.swi=4;
    tempRoundview.data=pullData;
    return view;
}
```

（2）把收到的温度数据设置到视图中。实现代码如下：

```
public void handleMessage(Message msg){
final TempFragment activity=mActivityReference.get();
if (activity!=null){                             //判断函数是否为空
switch (msg.what){
    cAndroid Studioe 1:
    double value=(Double) msg.obj;
    String string=String.valueOf(value);
    int idx=string.lAndroid StudiotIndexOf(".");
    String strNum=string.substring(0, idx);
    int num=Integer.valueOf(strNum);
    if (activity.oldHear!=num){
    activity.tempRoundbar.setCompany(activity.getResources().getString(R.string.centigrade));
    activity.tempRoundbar.setProgress(value);
    activity.pullData.add(num);
    if(activity.pullData.size()>activity.tempRoundview.MaxDatAndroid Studioize){
        activity.pullData.remove(0);
    }
    activity.tempRoundview.data=activity.pullData;
    activity.tempRoundview.swi=4;
    activity.tempRoundview.company=activity.getResources().getString(R.string.centigrade);
```

```
        activity.tempRoundview.company=activity.getResources().getString(R.string.
        centigrade);
        activity.tempRoundview.handler.sendEmptyMessage(1);
        activity.oldHear=num;
    }
break;
```

3. 采集温度数据核心代码

```
DataL = RX_byte(ACK);                    //通过发送ACK将低字节的数据接收回来
DataH = RX_byte(ACK);                    //通过发送ACK将高字节的数据接收回来
Pec = RX_byte(NACK);                     //通过发送ACK将校验字节的数据接收回来
STOP_bit();                              //结束通信
arr[0] = SlaveAddress;
arr[1] = command;
arr[2] = SlaveAddress;                   //将数据载入校验计算数组
arr[3] = DataL;
arr[4] = DataH;
arr[5] = 0;
PecReg = PEC_calculation(arr);           //计算校验字节
while(PecReg!= Pec);                     //如果计算的校验字节与接收的校验不一致,那么重新执
                                         //行以下程序
* ((unsigned char * )(&data)) = DataH;
* ((unsigned char * )(&data) +1) = DataL ;      //data = DataH:DataL
return data;}
```

4. 摄氏度换算函数代码

```
static float CalcTemp(uint16_t value)
{
    float temp;
    temp= (value*0.02)-273.15;
    return temp;
}
```

IR 传感器包括串联的热电偶,冷接头放置在厚的芯片衬底上,热接头放置在薄膜上。薄膜加热(或是冷却)从而吸收并辐射 IR。热电堆的输出信号为 Vir(Ta,To) = A.(To4 − Ta 4),其中 To 是物体的绝对温度(开尔文),Ta 是传感器片绝对温度,A 是总体的敏感度。需要一个附加的传感器来测量芯片的温度。在测量完两个传感器输出后,对应的环境温度和物体温度被计

算出。计算是通过内部 DSP 产生数字输出并和测量温度成线性比例。

环境温度 Ta 传感器芯片温度通过 PTAT 元件测量。所有传感器的状态和数据处理都是片上操作的,线性的传感器温度 Ta 存于 RAM 中。计算所得的 Ta 分辨率为 0.02 ℃。传感器出厂校准范围为 -40 ~ +85 ℃。在 RAM 单元地址 6h 中 2D89h 代表 -40 ℃,45F3h 代表 +85 ℃。

将 RAM 内容转换为实际的 Ta 比较简单:$Ta[K] = Tareg \times 0.02$。

公式转为代码应用:$temp = (value * 0.02) - 273.15$。

摄氏度数值存入 IIC 缓冲区,实现代码如下:

```
uint8_t temperature_getData(uint8_t * buff)
{
    buff[0] = tempHigh;         //赋值给数组
    buff[1] = tempLow;
    return 2;
}
```

temperature_getData( )函数在 iic_irqHandler( )函数中调用,其中 temperature_getData( )函数返回的是自己的特定数据类型标识,数组指针返回的则是要传输的摄氏数据。

只有在主 IIC 器件询问数据时才会触发数据的发送,所以整个过程是被动的。

5.2 可穿戴脚底压力模块

5.2.1 电阻应变式传感器原理及应用

电阻应变式传感器是以电阻应变计为转换元件的电阻式传感器。电阻应变式传感器由弹性敏感元件、电阻应变计、补偿电阻和外壳组成,可根据具体测量要求设计成多种结构形式。弹性敏感元件受到所测量的力而产生变形,并使附着其上的电阻应变计一起变形。电阻应变计再将变形转换为电阻值的变化,从而可以测量拉力、压力、扭矩、位移、加速度和温度等多种物理量。

这里所使用的拉力传感器及压力传感器均为电阻应变式传感器。应变片如图 5-14 所示。

优点:常用的电阻应变式传感器有应变式测力传感器、应变式压力传感器、应变式扭矩传感器(见转矩传感器)、应变式位移传感器(见位移传感器)、应变式加速度传感器(见加速度计)和测温应变计等。电阻应变式传感器的优点是精度高、测量范围广、寿命长、结构简单、频响特性好,能在恶劣条件下工作,易于实现小型化、整体化和品种多样化等。

图 5-14 应变片

缺点:对于大应变有较大的非线性、输出信号较弱,但可采取一定的补偿措施。因此,它广泛应用于自动测试和控制技术中。

5.2.2 拉力传感器与压力传感器

1. 拉力传感器的构造

拉力传感器隶属于称重传感器系列,是一种将物理信号转变为可测量的电信号输出的装置,它使用两个拉力传递部分传力,在其结构中含有力敏器件和两个拉力传递部分,在力敏器件中含有压电片、压电片垫片,后者含有基板部分和边缘传力部分,其特征是使两个拉力传递部分的两端分别固定在一起,用两端之间的横向作用面将力敏器件夹紧,压电片垫片在一侧压在压电片的中心区域,基板部分位于压电片另一侧与边缘传力部分之间并紧贴压电片,其用途之一是制成钩秤以取代杆秤。常见的拉力传感器如图 5-15 所示。

图 5-15 常见的拉力传感器

2. 拉力传感器使用注意事项

(1)设计加载装置及安装时应保证加载力的作用线与拉力传感器受力轴线重合,使倾斜负荷和偏心负荷的影响减至最小。

(2)在水平调整方面。如果使用的是单只拉力传感器,其底座的安装平面要使用水平仪调整直到水平;如果是多个传感器同时测量的情况,那么它们底座的安装面要尽量保持在一个水平面上,这样做的目的主要是为了保证每个传感器所承受的力量基本一致。

(3)按照其说明中拉力传感器的量程选定来确定所用传感器的额定载荷。

(4)传感器的底座安装面应尽可能平整和清洁,没有任何油污或者胶膜等存在。安装底座本身应具备足够的强度和刚性,通常要求高于传感器本身的强度和刚度。

3. 压力传感器的构造

压力传感器机电部分利用的是压电晶体的正压电效应。其原理是某些晶体(如人工极化陶瓷、压电石英晶体等,不同的压电材料具有不同的压电系数,一般都可以在压电材料性能表中查到)在一定方向的外力作用下或承受变形时,它的晶体面或极化面上将有电荷产生,这种从机械能(力,变形)到电能(电荷,电场)的变换称为正压电效应。压力传感器流程图如图 5-16 所示。

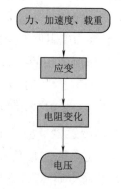

图 5-16 压力传感器流程图

4. 压力传感器的测量误差

造成传感器误差的因素有很多,下面说明一下 4 个无法避免的误差,这是压力传感器的初始误差。

(1)偏移量误差:由于压力传感器在整个压力范围内垂直偏移保持恒定,因此将产生偏移量误差。

(2)灵敏度误差:产生误差大小与压力成正比。如果设备的灵敏度高于典型值,灵敏度误差将是压力的递增函数。如果灵敏度低于典型值,那么灵敏度误差将是压力的递减函数。该误差的产生原因在于扩散过程的变化。

(3)线性误差:这是一个对压力传感器初始误差影响较小的因素,该误差的产生原因在于应变片的物理非线性,但对于带放大器的传感器,还应包括放大器的非线性。线性误差曲线可以是凹形曲线,也可以是凸形曲线。

(4)滞后误差:在大多数情形中,压力传感器的滞后误差完全可以忽略不计,一般只需在压力变化很大的情形中考虑滞后误差。

5.2.3　力敏传感器应用电路

1. HX711 转换芯片简介

HX711 采用了海芯科技集成电路专利技术,是一款专为高精度电子秤而设计的 24 位 A/D 转换器芯片。与同类型其他芯片相比,该芯片集成了包括稳压电源、片内时钟振荡器等其他同类型芯片所需要的外围电路,具有集成度高、响应速度快、抗干扰性强等优点。降低了电子秤的整机成本,提高了整机的性能和可靠性。HX711 电路板的引脚图如图 5 - 17 所示。

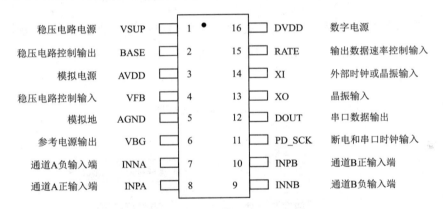

图 5 - 17　HX711 电路板引脚图

该芯片与后端 MCU 芯片的接口和编程非常简单,所有控制信号由引脚驱动,无须对芯片内部的寄存器编程。输入选择开关可任意选取通道 A 或通道 B,与其内部的低噪声可编程放大器相连。通道 A 的可编程增益为 128 或 64,对应的满额度差分输入信号幅值分别为 ±20 mV 或 ±40 mV。通道 B 则为固定的 32 增益,用于系统参数检测。芯片内提供的稳压电源可以直接向外部传感器和芯片内的 A/D 转换器提供电源,系统板上无须另外的模拟电源。芯片内的时钟振荡器不需要任何外接器件。上电自动复位功能简化了开机的初始化过程。

2. HX711 电路参考应用

在本项目中,针对力敏传感器电路主要以 HX711 的计价秤的参考电路为依据,其电路图如图 5 - 18 所示。该方案使用内部时钟振荡器(XI = 0),10 Hz 的输出数据速率(RATE = 0)。

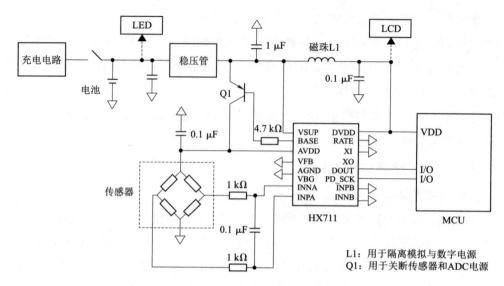

图 5-18　HX711 芯片应用于计价秤的参考电路图

电源(2.7~5.5 V)直接取用与 MCU 芯片相同的供电电源。通道 A 与传感器相连,通道 B 通过片外分压电阻(未在图一中显示)与电池相连,用于检测电池电压。

5.2.4　拉力与压力程序解析

根据脚底压力项目设计原则,其程序的软件流程图如图 5-19 所示。

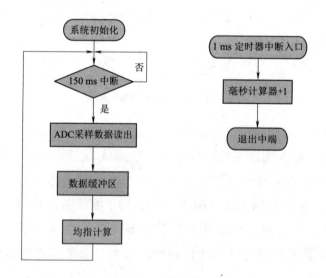

图 5-19　软件流程图

流程步骤如下:

(1)单片机通过和 HX711 进行串行通信,将 HX711 外接传感器的压变数值读回。

(2)压力和拉力的应用方式相差无几,区别的地方在于拉力和压力的压变系数不一样,应力变化的方向也不一样,所以用了两个不同的函数进行区分。

(3)根据 150 ms 的周期间隔把数值转换为重量单位并载入到 IIC 发送缓冲区及在 OLED 上显示。

- ADC 采样数据。实现代码如下:

```
uint32_t forceReadCount(void)
{
uint32_t Count;
uint8_t i;
ADSKOut(OUT_LOW);                          //使能 AD(PD_SCK 置低)
Count=0;
while (ADDOIn());                          //AD 转换未结束则等待,否则开始读取
for (i=0; i<24; i++){
    ADSKOut(OUT_HIGH);                     //PD_SCK 置高(发送脉冲)
    Count=Count <<1;                       //下降沿来时变量 Count 左移一位,右侧补零
    ADSKOut(OUT_LOW);                      //PD_SCK 置低
    if (ADDOIn()){                         //判断接收电平信号
        Count++;
    }
ADSKOut(OUT_HIGH);
Count = Count ^ 0x800000;                  //第 25 个脉冲下降沿来时,转换数据
ADSKOut(OUT_LOW);
return(Count);
}
```

- 数值转换及显示。实现代码如下:

```
if(flag==1){
    flag=0;
    f_timedelay=5000;                      //暂缓执行时间赋值,暂缓 5S 执行
}
readData=forceReadCount();                 //读出压变数据
readBuff[readBuffCount++] = readData;      //将压变数据放入滤波缓冲区
if (readBuffCount>=FORCE_READ_DATA_BUFF_SIZE){  //判断缓冲区是否已满
    bubble_sort(readBuff,readBuffCount);   //数据冒泡排序
    for(i=0;i<readBuffCount-2;i++) readData+=readBuff[i+1]; //放弃头尾两个数据
    readData=readData/(readBuffCount-2);   //针对中间数据进行累加
    readBuffCount=0;
```

智能可穿戴设备的设计与实现

```
        if(f_timedelay==0){                                    //周期执行判断
            if (0==correctValue) {
                correctValue=readData;                         //赋予初次校准值
            }
            else{
                #if(BLE_BOARD==BLE_BOARD_TYPE_PULL)//根据预编译条件决定拉力或压力转换显示函数
                    forcePullConverter(readData-correctValue);
                #elif(BLE_BOARD==BLE_BOARD_TYPE_PRESSURE)
                    forcePressureConverter(correctValue-readData);
                #endif
            }
            f_timedelay=DELAY_COUNT;    //周期计数变量重新赋值,计数变量在定时器中断执行自减
        }
```

- 重量数值存入 IIC 缓冲区。实现代码如下：

```
uint8_t correct_getData(uint8_t* buff)
{
    buff[0]=weightHigh;
    buff[1]=weightLow;
    return 2;
}
```

correct_getData( )函数在 iic_irqHandler( )函数中调用,其中 correct_getData( )函数返回的是自己的数据长度,数组指针返回的则是要传输的总量。

只有在主 IIC 器件询问数据时才会触发数据的发送,所以整个过程是被动的。

5.2.5　手机 APP 软件的开发和功能

1. 压力模块协议体具体定义(见表 5－2)

表 5－2　压力模块协议体

设 备 号	有效数据长度	有效数据区	
		压力整数部分	压力小数部分
0x07	0x02	－100～100	0～255

数据解析:在第 4 章的章节中说到通过 onCharacteristicChanged( )函数回调拿到可穿戴模块的数据 20 个 byte 的数组。通过判断 byte 数组的第一个字节可以知道当前返回的模块设备号;判断 byte 数组的第二个字节可以知道当前数组的有效字节长度;判断 byte 数组的第三个字节

可以知道压力的整数数值；判断 byte 数组的第四个字节可以知道压力的小数数值。实现代码如下：

```
cAndroid Studioe 0x07:
if (value[1]==0x02){
    int integer=value[2]>=0? value[2] : value[2]+256;
    int decimal=value[3]>=0? value[3] : value[3]+256;
    double dou=Double.valueOf(String.valueOf(integer) +"."+String.valueOf(decimal));
if(activity.weData != null){                          //获取模块设备号
    activity.weData.setPressure(String.valueOf(dou));
}
if(activity.pressureFragment != null){
    if(activity.currentFragment==activity.pressureFragment){
        if (activity.pressureFragment.handler!=null){
            if (dou>=0){
                Message message=new Message();         //获取新的信息
                message.what=1;
                message.obj=dou;
                activity.pressureFragment.handler.sendMessage(message);
            }
        }
    }
} else if (activity.currentFragment==activity.humanFallFragment){
activity.humanFallFragment.number= integer;
}
}
break;
```

因为 byte 的一个字节是 8 位，范围为 $-128 \sim 127$，即 $-2^7 \sim 2^7-1$，所以在判断压力数值时，需要判断数值是否大于零。如果小于零则需要加上 256 将其转换为正整数。

2. 自定义视图

自定义 RoundProgressBar 和 RoundView 显示在 PressureFragment 中。

自定义 RoundProgressBar 的内容包括：

- 自定义 View 的属性。
- 在 View 的构造方法中获得自定义的属性。
- 重写 onMesure（onMesure 不一定是必需的）。
- 重写 onDraw。

(1) 自定义 RoundProgressBar：

● 在 res/values/下创建 attrs.xml；自定义 view 的属性，具体实现代码如下：

```xml
<?xml version="1.0" encoding="UTF-8"?>
<resources>
<declare-styleable name="RoundProgressBar">        //自定义 view 的属性
<attr name="roundColor" format="color"/>
<attr name="roundProgressColor" format="color"/>
<attr name="roundWidth" format="dimension"/>
<attr name="textColor" format="color"/>
<attr name="textSize" format="dimension"/>
<attr name="max" format="integer"/>
<attr name="textIsDisplayable" format="boolean"/>
<attr name="style">
<enum name="STROKE" value="0"/>
<enum name="FILL" value="1"/>
</attr>
</declare-styleable>
</resources>
```

● 在布局 fragment_pressure.xml 中声明自定义 View。实现代码如下：

```xml
<activity.lsen.wearabledevice.tool.RoundProgressBar
    android:id="@+id/pressure_roundbar"
    android:layout_width="200dp"
    android:layout_height="200dp"
    android:layout_gravity="center"            //设置中心位置
    android:layout_marginTop="10dp"
    app:roundColor="@color/round_bar"
    app:roundWidth="10dp"
    app:textColor="@color/text_color"          //设置文本颜色
    app:textSize="@dimen/text_title" />
```

● 在 View 的构造方法中获得自定义的属性。实现代码如下：

```java
public RoundProgressBar(Context context, AttributeSet attrs, int defStyle){
    super(context, attrs, defStyle);
    paint=new Paint();
    TypedArray mTypedArray=context.obtainStyledAttributes(attrs,
```

```
        R.styleable.RoundProgressBar);
roundColor=mTypedArray.getColor(
        R.styleable.RoundProgressBar_roundColor, Color.RED);
roundProgressColor=mTypedArray.getColor(
        R.styleable.RoundProgressBar_roundProgressColor, Color.GREEN);
textColor=mTypedArray.getColor(
        R.styleable.RoundProgressBar_textColor, Color.GREEN);
textSize=mTypedArray.getDimension(
        R.styleable.RoundProgressBar_textSize, 15);
roundWidth=mTypedArray.getDimension(
        R.styleable.RoundProgressBar_roundWidth, 40);
max=mTypedArray.getInteger(R.styleable.RoundProgressBar_max, 140);
textIsDisplayable=mTypedArray.getBoolean(
        R.styleable.RoundProgressBar_textIsDisplayable, true);
style=mTypedArray.getInt(R.styleable.RoundProgressBar_style, 0);
mTypedArray.recycle();
}
```

- 重写 OnDraw。实现代码如下:

```
@Override
protected void onDraw(CanvAndroid Studio canvAndroid Studio){
    super.onDraw(canvAndroid Studio);
    int centre=getWidth() / 2;
    int radius= (int) (centre- roundWidth / 2);
    paint.setColor(roundColor);
    paint.setStyle(Paint.Style.STROKE);
    paint.setStrokeWidth(roundWidth);
    paint.setAntiAliAndroid Studio(true);
    canvAndroid Studio.drawCircle(centre, centre, radius, paint);
    paint.setStrokeWidth(0);
    paint.setColor(textColor);
    paint.setTextSize(textSize);
    paint.setTypeface(Typeface.DEFAULT_BOLD);
    double percent= (double) (((float) progress / (float) max)*140);
    percent= formatDouble(percent);
    float textWidth=paint.meAndroid StudioureText(percent +company);
```

```
            if (textIsDisplayable && percent !=0 && style==STROKE) {
                canvAndroid Studio.drawText(percent+company, centre-textWidth / 2, centre
                    +textSize / 2, paint);
            } else {
                canvAndroid Studio.drawText(percent+company, centre-textWidth / 2, centre
                    +textSize / 2, paint);
            }
            paint.setStrokeWidth(roundWidth);
            paint.setColor(textColor);
            RectF oval=new RectF(centre-radius, centre-radius, centre
                +radius, centre+radius);
            switch (style){
                cAndroid Studioe STROKE: {
                    paint.setStyle(Paint.Style.STROKE);
                    canvAndroid Studio.drawArc(oval, 0,(float)(360* progress / max), false,
                        paint);
                    break;
                }
                cAndroid Studioe FILL: {
                    paint.setStyle(Paint.Style.FILL_AND_STROKE);
                    if (progress !=0)
                        canvAndroid Studio.drawArc(oval, 0, (float) (360 * progress / max),
                            true, paint);
                    break;
                }
            }
        }
```

● RoundProgressBar 效果如图 5-20 所示。

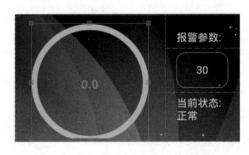

图 5-20　RoundProgressBar 效果

- 在 RoundProgressBar 中设置数值的接口。实现代码如下：

```
public synchronized void setProgress(double progress){
    if (progress >max){
        progress =max;
    }
    if (progress <=max){
        this.progress=progress;
        postInvalidate();
    }
}
```

（2）自定义 PressureView：

该视图未做屏幕适配处理,如果需要,可以自己尝试修改。实现代码如下：

- 绘制曲线边线的 XY 轴。实现代码如下：

```
@Override
protected void onDraw(CanvAndroid Studio canvAndroid Studio){
    super.onDraw(canvAndroid Studio);
    if (swit==0){
        YScale=4;
        YLabel= new String[YLength / YScale];
        for (int i=0; i<YLabel.length; i++) {
            YLabel[i]= (i) +"";
        }
    } else {
        xy=180;
        YScale=3;
        YLabel= new String[YLength / YScale];
        for (int i=0; i<YLabel.length; i++){
            YLabel[i]= (i-60) +"";
        }
    }
    Paint paint= new Paint();
    paint.setStyle(Paint.Style.STROKE);
    paint.setAntiAliAndroid Studio(true);
    paint.setTextSize(25);
    paint.setColor(Color.WHITE);
```

```
Paint paint2=new Paint();
paint2.setStyle(Paint.Style.STROKE);
paint2.setAntiAliAndroid Studio(true);
paint2.setColor(Color.RED);
canvAndroid Studio.drawLine(XPoint, YPoint-YLength, XPoint, YPoint, paint);
```

- 不同的模块有不同的曲线要求,所以根据模块来绘制。实现代码如下:

```
if (swit==0) {
    for (int i = 0; i * YScale < YLength; i++) {
        if (i==0 ||i==10 ||i==20 ||i==30 ||i==40 ||i==50
            ||i==60 ||i==70 ||i==80 ||i==90 ||i==100
            ||i==110){
            canvAndroid Studio.drawLine(XPoint, YPoint-i*YScale, XPoint+5, YPoint
                -i*YScale, paint);
            canvAndroid Studio.drawText(YLabel[i], XPoint-50,
                YPoint-i*YScale+3, paint);
        }
    }
} else{
    YScale=3;
    for(int i=0; i*YScale<YLength; i++){
        if (i==0 ||i==20 ||i==40 ||i==60 ||i==80 ||i==100
            ||i==120 ||i==140 ||i==160 ||i==180 ||i==200
            ||i==220){
            canvAndroid Studio.drawLine(XPoint, YPoint-i*YScale, XPoint+5, YPoint
                - i * YScale, paint);
            canvAndroid Studio.drawText(YLabel[i], XPoint-50,
                YPoint-i*YScale+3, paint);
        }
    }
}
canvAndroid Studio.drawText("单位:kg", XPoint-30, YPoint-YLength-20, paint);
canvAndroid Studio.drawLine(XPoint, YPoint-YLength, XPoint+XLength, YPoint
    -YLength, paint2);
canvAndroid Studio.drawLine(XPoint, YPoint, XPoint + XLength, YPoint, paint);
```

- 绘制曲线。实现代码如下：

```
if (data.size() >1){
if(swit==0){
    for (int i=1; i<data.size(); i++){
    canvAndroid Studio.drawLine((float) (XPoint+ (i-1)*XScale),
    (float) (YPoint-data.get(i-1)*YScale) ,
    (float) (XPoint+i*XScale),
    (float) (YPoint-data.get(i)*YScale) , paint);
    }
}else{
    for (int i=1; i<data.size(); i++){
    canvAndroid Studio.drawLine((float) (XPoint+ (i-1)*XScale),
    (float) (YPoint-data.get(i-1)*YScale) -xy,
    (float) (XPoint+i*XScale),
    (float) (YPoint-data.get(i)*YScale)-xy, paint);
    }
  }
}
```

因为曲线需要动态的显示,所以需要不断地更新数据调用 invalidate( )来重写 OnDraw( )函数。实现代码如下：

```
private static clAndroid Studios HearHandler extends Handler{
    private WeakReference<PressureView> mActivityReference;
    //不断更新数据
    HearHandler(PressureView activity){
        mActivityReference=new WeakReference<>(activity);
    }
    //重写 OnDraw
    @Override
    public void handleMessage(Message msg){
        final PressureView activity=mActivityReference.get();
        if(activity != null) {
            if(msg.what==1){
                if(activity.data.size() >=activity.MaxDatAndroid Studioize){
                    activity.data.remove(0);   //删除数据
                }
```

```
            activity.invalidate();
        }
    }
}
```

- 在布局 fragment_pressure.xml 中声明自定义 View。实现代码如下：

```
<activity.lsen.wearabledevice.view.PressureView
    android:id = "@ + id/pressure_roundview"
    android:layout_width = "match_parent"
    android:layout_height = "0dp"
    android:layout_gravity = "center"
    android:layout_marginTop = "10dp"
    android:layout_weight = "2" />
```

最终效果图 5 - 21 所示。

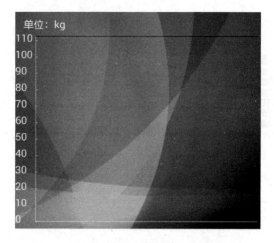

图 5 - 21　曲线图

3. PressureFragment 中使用 RoundProgressBar 和 PressureView

（1）在 onCreateView( ) 中使用视图。实现代码如下：

```
public View onCreateView(LayoutInflater inflater, @Nullable ViewGroup container,
     @Nullable Bundle savedInstanceState) {
  View view= inflater.inflate(R.layout.fragment_pressure, container, false);
  unbinder= ButterKnife.bind(this, view);   //建立视图 PressureView
  pressureRoundview.swit=0;
```

```
pressureRoundview.data=pullData;
alarm=Double.valueOf(pressureNumberEt.getText().toString());
```

(2)把收到的压力数据设置到视图中。实现代码如下:

```
@Override
public void handleMessage(Message msg) {
final PressureFragment activity=mActivityReference.get();
if (activity != null) {
    switch (msg.what) {
    cAndroid Studioe 1:
    double value= (double) msg.obj;
    String string= String.valueOf(value);
    int idx= string.lAndroid StudiotIndexOf(".");
    String strNum= string.substring(0, idx);
    int num= Integer.valueOf(strNum);
if (activity.oldHear != num){
if (value >= activity.alarm){
    activity.pressureStateTv.setTextColor(0xffffff00);
    activity.pressureStateTv.setText(activity.getResources().getString(R.string.
    uv_state2));
}else{
    activity.pressureStateTv.setTextColor(0xffffffff);     //设置字体颜色
    activity.pressureStateTv.setText(activity.getResources().getString(R.string.
    uv_state1));
}
activity.pullData.add(num);
if(activity.pullData.size()>activity.pressureRoundview.MaxDatAndroid Studioize){
    activity.pullData.remove(0);                           //删除相关线条
}
    activity.pressureRoundview.data=activity.pullData;
    activity.pressureRoundbar.setCompany(activity.getResources().getString(R.string.
    cattle));
    activity.pressureRoundbar.setProgress(value);
    activity.pressureRoundview.swit=0;
    activity.pressureRoundview.handler.sendEmptyMessage(1);
    activity.oldHear = num;
  }
break;
```

- 根据输入的报警参数,设置压力报警。当输入结束后,判断报警参数是否在压力范围内,如果在则输入成功。实现代码如下:

```
alarm= Double.valueOf(pressureNumberEt.getText().toString());
pressureNumberEt.addTextChangedListener(new TextWatcher(){
    @Override
    public void beforeTextChanged(CharSequence s, int start, int count, int after){
        pressureRoundview.swit=0;
        pressureRoundview.data=pullData;
    }
    @Override
    public void onTextChanged(CharSequence s, int start, int before, int count){
        forcePullConverter(readData-correctValue);
    }
    @Override
    public void afterTextChanged(Editable s){
        String numStr=pressureNumberEt.getText().toString();
        if (!numStr.isEmpty()){   //判断报警参数是否在压力范围内
            double num= Double.valueOf(numStr);
            if (num <-100 ||num >120){
                handler.sendEmptyMessage(2);
            } else{
                alarm= num;
            }
        }
    }
}
```

- 压力参数报警判断。当压力值大于或等于报警参数时,则提示报警。实现代码如下:

```
if (value >= activity.alarm){
    activity.pressureStateTv.setTextColor(0xffffff00);   //提示报警
    activity.pressureStateTv.setText(activity.getResources().getString(R.string.
        uv_state2));
}else{
    activity.pressureStateTv.setTextColor(0xffffffff);   //提示报警
    activity.pressureStateTv.setText(activity.getResources().getString(R.
        string.uv_state1));
}
```

5.3 心率计数模块

5.3.1 心率采集信息价值

心率是指正常人安静状态下每分钟心跳的次数,也称安静心率,一般为 60～100 次/min,可因年龄、性别或其他生理因素产生个体差异。一般来说,年龄越小,心率越快,老年人心跳比年轻人慢,女性的心率比同龄男性快,这些都是正常的生理现象。安静状态下,成人正常心率为 60～100 次/min,理想心率应为 55～70 次/min(运动员的心率较普通成人偏慢,一般为 50 次/min 左右)。图 5-22 所示为心率图片。

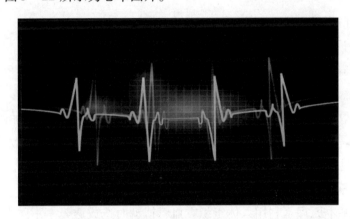

图 5-22 心率图片

心率采集常见问题如下:

1. 心率过速

成人安静时心率超过 100 次/min(一般不超过 160 次/min),称为窦性心动过速,常见于兴奋、激动、吸烟、饮酒、喝浓茶或咖啡后,或见于感染、发热、休克、贫血、缺氧、甲亢、心力衰竭等病理状态下,或见于应用阿托品、肾上腺素、麻黄素等药物后。

2. 心率过缓

成人安静时心率低于 60 次/min(一般在 45 次/min 以上),称为窦性心动过缓,可见于长期从事重体力劳动的健康人和运动员;或见于甲状腺机能低下、颅内压增高、阻塞性黄疸以及洋地黄、奎尼丁或心得安类药物过量。如果心率低于 40 次/min,应考虑有病态窦房结综合征、房室传导阻滞等情况。如果脉搏强弱不等、不齐且脉率少于心率,应考虑心房纤颤。

3. 窦性心动过缓

很多人都会有窦性心动过缓伴不齐,对于多数人来说是正常的,不必过于担心。窦性心动过缓是指心率低于 60 次/min 的人,是否会出现此症状,与其心跳过缓的频率和引起心跳过缓的原因有关。在安静状态下,成年人的心率若在 50～60 次/min 之间一般不会出现明显症状。

尤其是一些训练有素的运动员以及长期从事体力劳动的人，在安静状态下即使其心率在 40 次/min 左右也不会出现明显症状。但是，一般人的心率若在 40~50 次/min 之间，就会出现胸闷、乏力、头晕等症状；若其心率降至 35~40 次/min 则会发生血流动力学改变，使心脑器官的供血受到影响，从而出现胸部闷痛、头晕、晕厥甚至猝死。如果自我感觉没有任何不适，不用去理会心电图所说的"窦性心动过缓伴不齐"；如果出现胸闷、乏力、头晕等不适症状，应立即到医院进一步检查（如动态心电图、心脏彩超等检查），了解心动过缓的病因；如果心跳过慢，可以通过安装心脏起搏器缓解症状，改善预后。

5.3.2 非接触式体温传感器原理

1. 心率测量方法

传统的心率脉搏测量方法主要有 3 种：

（1）从心电信号中提取。

（2）从测量血压时压力传感器测到的波动来计算脉率。

（3）光电容积法。

前两种方法提取信号都会限制监测对象的活动，如果长时间使用会增加监测对象的生理和心理上的不舒适感。而光电容积法脉搏测量作为监护测量中最普遍的方法之一，具有方法简单、佩戴方便、可靠性高等特点。

2. 光电容积法原理

光电容积法的基本原理是利用人体组织在血管搏动时造成透光率不同来进行脉搏测量的。其使用的传感器由光源和光电变换器两部分组成，通过绑带或夹子固定在病人的手腕上。当光束透过人体外周血管，由于动脉搏动充血容积变化导致这束光的透光率发生改变，此时由光电变换器接收经人体组织反射的光线，转变为电信号并将其放大和输出。由于脉搏是随心脏的搏动而周期性变化的信号，动脉血管容积也周期性地变化，因此光电变换器的电信号变化周期就是脉搏率。光电容积法原理如图 5-23 所示。

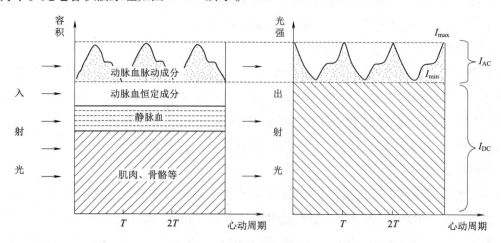

图 5-23 光电容积法原理

根据相关文献和实验结果,560 nm 的光波可以反映皮肤浅部微动脉信息,适合用来提取脉搏信号。该传感器主动发射峰值波长为 515 nm 的绿光 LED,再通过光接收器拾取反射光谱,由于脉搏信号的频带一般在 0.05~200 Hz 之间,信号幅度均很小,一般在毫伏级水平,容易受到各种信号干扰。在感受器后面使用了低通滤波器和运放构成的放大器,将信号放大了数百倍后,信号可以很好地被单片机的 A/D 转换器采集到。心率传感器的工作示意图如图 5-24 所示。

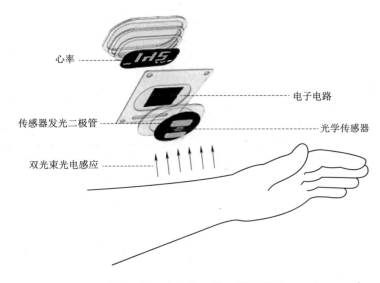

图 5-24 心率传感器工作示意图

5.3.3 心率传感器电路解析

本模块使用 SON1303 光电式心率传感器,可放置于人体各部位测试人体心率和脉搏。

(1) SON1303 采用的反射式光电传感器使测量方式更加自由,应用范围遍及可佩戴式电子产品以及新式测试方法的脉搏测量仪器,能扩大脉搏测量配套设备的应用范围。

(2) 内部集成高科技纳米涂层环境光检测传感器,过滤不需要的光源,减少由其他光源干扰的误判动作,准确度高。

(3) SON1303 采用了 570 nm 发光波长的绿光,与红外光相比反射率更高,测量感度更高,同时提高了波长比特性,使用了最适合测量脉搏用的发光波长。

心率传感模块,其中集成 SON1303 和 SON3130。SON1303 作为心率传感芯片,配合 SON3130 使用;SON3130 是高阻型运算放大器,可将 SON1303 采集到的信号进行放大输出。其电路如图 5-25 所示。

5.3.4 心率传感代码解析

1. MCU 嵌入式系统功能

通过 ADC 采集传感器输出的脉搏的模拟信号,并通过数字滤波算法去除干扰信号,然后将有效脉搏信号通过蓝牙模块传输到 APP 中。

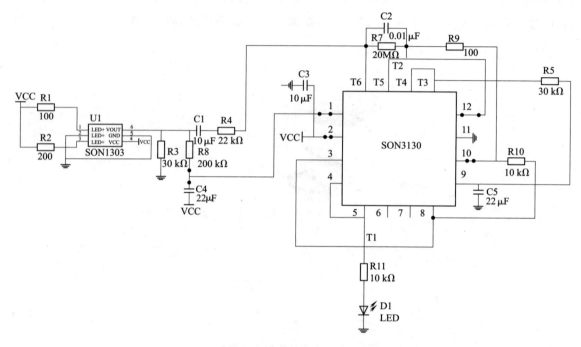

图 5-25　SON1303 电路图

脱机工作时通过波形识别算法,识别出每一个有效的脉搏脉冲信号,换算出人体准实时心率,及心率的变化;以供与 APP 连接上后可以查询历史心率变化。采集到的模拟信号如图 5-26 所示。

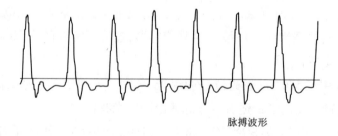

图 5-26　心率波形图

2. 手机 APP 软件功能

收集 MCU 发送来的脉搏模拟信号,并将其存入到数据库中,形成一系列人体脉搏的历史记录。然后,可以通过调取这些记录描绘出人的历史脉搏曲线。

通过波形处理算法,识别出每一个有效的脉搏信号,并通过每两个脉搏之前的时间间隔计算出瞬时的心率。其软件流程图如图 5-27 所示。

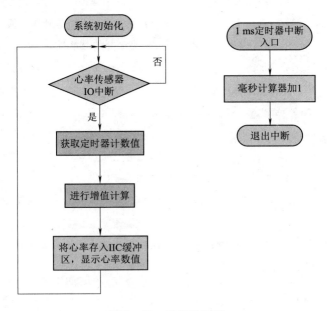

图 5-27 软件流程图

SON3130 芯片内部集成有施密特触发器,可以将脉搏流形整形成矩形,简化单片软件的结构。

单片机通过外部中断和定时器测量脉搏的周期,从而计算出人体的心率。

计算过程中考虑到测量过程有干扰的存在,所以中间使用了均值滤波算法将其中的干扰剔除,最终获得稳定的心率值。

获取定时器计数值代码:

```
void heart_appHandler(void)
{
    heartSkip=TRUE;
    time=getSystemMsTime();
}
```

获取定时器计数值的函数被放在检测脉冲波形中断引脚的中断服务函数里,中断服务函数为 INTERRUPT_HANDLER(EXTI5_IRQHandler, 13)。

只有当外部中断引脚被脉冲跳变沿触发时中断服务函数内的 heart_appHandler() 函数才会进行定时器计数值的获取。

(1)心率数值计算实现代码如下:

```
if(FALSE==heartSkip) return;        //如果外部中断没有被触发,则直接返回
    newTime=time;
```

```
    tmpRate=60000 / (newTime-oldTime);                    // 计算本周期心率值
    oldTime=newTime;                       // 保存本次中断的计数器值,供下次中断计数周期
    if (tmpRate >0 && tmpRate <140){       // 去除异常心率值
        heartRateBuff[heartCnt++] = (uint8_t)tmpRate;     //将新心率数值存入均值滤波
                                                            缓冲区

        heartCnt %= HEART_MARK_TIME_BUFF_SIZE;             //滤波缓冲区数组咬尾判断
        rateSum=0;
        //累加数值清零
        for (i=0; i<HEART_MARK_TIME_BUFF_SIZE;i++){
            rateSum += heartRateBuff[i];                   //心率数值累加计算
        }
        heartRate= rateSum / HEART_MARK_TIME_BUFF_SIZE;    //心率均值计算
        char strBuff[17] ="";
        uint8_t len;
        intNumToStr(heartRate, & strBuff[strlen(strBuff)]); //格式化打印
        strcat(strBuff, "BPM   ");                          //格式化打印
        len= strlen(strBuff);
        //格式化打印
        while ((len++) <16){
            strcat(strBuff, "");
                                                            //格式化打印
        }
        OLED_ShowString(0,4,strBuff);                       //根据格式化打印内容进行显示
    }
```

(2)心率数值传入 IIC 缓冲区。实现代码如下:

```
uint8_t heart_getData(uint8_t * buff)
{
    buff[0]=heartRate;
    return 0x01;
}
```

heart_getData( )函数在 iic_irqHandler( )函数中调用,其中 heart_getData( )函数返回的是自己的数据长度,数组指针返回的则是要传输的心率数值。

只有在主 IIC 器件询问数据时才会触发数据的发送,所以整个过程是被动的。

5.3.5 手机 APP 软件的开发和功能

1. 心率模块协议体具体定义（表 5-3）

表 5-3 心率模块协议体

设 备 号	长 度	有效数据区
		心率
0x01	0x01	0~140:已测的心率； 小于 0:心率计故障

心率解释:心率是指正常人安静状态下每分钟心跳的次数,也称安静心率,一般为 60~100 次/分,可因年龄、性别或其他生理因素产生个体差异。一般来说,年龄越小,心率越快,老年人心跳比年轻人慢,女性的心率比同龄男性快,这些都是正常的生理现象。

数据解析:在第 4 章讲到通过 onCharacteristicChanged()函数回调拿到可穿戴模块的数据 20 个 byte 的数组。通过判断 byte 数组的第一个字节可以知道当前返回的模块设备号;判断 byte 数组的第二个字节可以知道当前数组的有效字节长度;判断 byte 数组的第三个字节可以知道心率的数值。实现代码如下:

```
byte[] value=activity.value;
switch (value[0]) {
    cAndroid Studioe 0x01:
    if (value[1]==0x01){
    int va=value[2] >=0 ? value[2] : (value[2] +256);
    if (activity.weData!=null){
    activity.weData.setHear(String.valueOf(va));
    }
        if (activity.heartFragment!=null) {    //判断 byte 数组
        if (activity.currentFragment== activity.heartFragment) {
        if (activity.heartFragment.hearHandler!=null) {
        Message message= new Message();
        message.what=1;
        message.obj=va;
        activity.heartFragment.hearHandler.sendMessage(message);
        }
        }
        }
        if (activity.motionFragment!=null){    //判断 byte 数组
        if (activity.currentFragment== activity.motionFragment) {
        if (activity.motionFragment.handler!=null){
        Message message= new Message();
        message.what=3;
        message.obj=va;
        activity.motionFragment.handler.sendMessage(message);
```

```
            }
        }
    }
}
break;
```

因为 byte 的一个字节是 8 位,范围为: $-128 \sim 127$,即 $-2^7 \sim 2^7-1$,所以在判断心率数值时,需要判断数值是否大于零。如果小于零,则需要加上 256 将其转换为正整数。

2. 自定义视图

自定义 RoundProgressBar 和自定义 RoundView 显示在 HeartFragment 中。

自定义 RoundProgressBar 的内容包括:

- 自定义 View 的属性。
- 在 View 的构造方法中获得我们自定义的属性。
- 重写 onMesure(onMesure 不一定是必需的)。
- 重写 onDraw。

(1) 自定义 RoundProgressBar:

- 在 res/values/下创建 attrs.xml;自定义 view 的属性。实现代码如下:

```xml
<?xml version="1.0" encoding="UTF-8"?>
<resources>
<declare-styleable name="RoundProgressBar">
<attr name="roundColor" format="color"/>
<attr name="roundProgressColor" format="color"/>
<attr name="roundWidth" format="dimension"/>
<attr name="textColor" format="color" />
<attr name="textSize" format="dimension" />
<attr name="max" format="integer"/>
<attr name="textIsDisplayable" format="boolean"/>
<attr name="style">
<enum name="STROKE" value="0"/>
<enum name="FILL" value="1"/>
</attr>
</declare-styleable>
</resources>
```

- 在布局 fragment_hear.xml 中声明自定义 View。实现代码如下:

```xml
<activity.lsen.wearabledevice.tool.RoundProgressBar
    android:id="@+id/hear_roundbar"
    android:layout_width="200dp"
```

```xml
        android:layout_height = "200dp"
        android:layout_gravity = "center"              //自定义 View
        android:layout_marginTop = "10dp"
        app:roundColor = "@color/round_bar"
        app:roundWidth = "10dp"
        app:textColor = "@color/text_color"            //app 颜色设置
        app:textSize = "@dimen/text_title" />
</LinearLayout>
```

- 在 View 的构造方法中获得自定义的属性。实现代码如下:

```java
public RoundProgressBar(Context context, AttributeSet attrs, int defStyle) {
    super(context, attrs, defStyle);
    paint=new Paint();
    TypedArray mTypedArray= context.obtainStyledAttributes(attrs,
        R.styleable.RoundProgressBar);
        R.styleable.RoundProgressBar_textIsDisplayable, true);
    style=mTypedArray.getInt(R.styleable.RoundProgressBar_style, 0);
    mTypedArray.recycle();
}
```

- 重写 OnDraw。实现代码如下:

```java
protected void onDraw(CanvAndroid Studio canvAndroid Studio) {
    super.onDraw(canvAndroid Studio);
    int centre= getWidth() / 2;
    int radius= (int) (centre- roundWidth / 2);        //数值计算
    paint.setColor(roundColor);
    paint.setStyle(Paint.Style.STROKE);
    paint.setStrokeWidth(roundWidth);
    paint.setAntiAliAndroid Studio(true);
    canvAndroid Studio.drawCircle(centre, centre, radius, paint);
    paint.setStrokeWidth(0);
    paint.setColor(textColor);
    paint.setTextSize(textSize);                       //字体大小设置
    paint.setTypeface(Typeface.DEFAULT_BOLD);
    double percent= (double) (((float) progress / (float) max)*140);
    percent= formatDouble(percent);
    float textWidth=paint.meAndroid StudioureText(percent+ company);
    if (textIsDisplayable && percent != 0 && style==STROKE) {
```

```
        canvAndroid Studio.drawText(percent+company, centre-textWidth / 2, centre
            +textSize / 2, paint);
    } else {
        canvAndroid Studio.drawText(percent+company, centre-textWidth / 2, centre
            +textSize / 2, paint);
    }
paint.setStrokeWidth(roundWidth);
paint.setColor(textColor);                    //设置线条颜色
RectF oval=new RectF(centre-radius, centre-radius, centre
        +radius, centre+radius);
switch (style){
    cAndroid Studioe STROKE: {
        paint.setStyle(Paint.Style.STROKE);
        canvAndroid Studio.drawArc(oval, 0, (float) (360 * progress / max), false,
            paint);
        break;
    }
    cAndroid Studioe FILL: {                  //判断是否画满屏幕
        paint.setStyle(Paint.Style.FILL_AND_STROKE);
        if (progress != 0)
            canvAndroid Studio.drawArc(oval, 0, (float) (360* progress / max),
                true, paint);
        break;
    }
}
}
```

● RoundProgressBar 效果如图 5-28 所示。

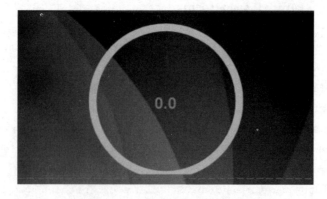

图 5-28 RoundProgressBar 效果图

- 在 RoundProgressBar 中设置数值的接口。实现代码如下:

```java
public synchronized void setProgress(double progress) {
    if (progress > max) {              //最小值
        progress=max;
    }
    if (progress <= max) {             //最大值
        this.progress=progress;
        postInvalidate();
    }
}
```

(2) 自定义 RoundView:

该视图未做屏幕适配处理,如果需要可以自己尝试修改。

- 绘制曲线边线的 XY 轴。实现代码如下:

```java
protected void onDraw(CanvAndroid Studio canvAndroid Studio){
    super.onDraw(canvAndroid Studio);
    MaxDatAndroid Studioize=XLength / XScale;
    YLabel= new String[YLength / YScale];
    if (swi==0 || swi==8){
        for (int i=0; i<YLabel.length; i++){
            YLabel[i] = (i) +"";
        }
    } else {
        for (int i=0; i<YLabel.length; i++) {
            YLabel[i] = (i-20) +"";
        }
    }
    Paint paint=new Paint();            //在视图列表中划线
    paint.setStyle(Paint.Style.STROKE);
    paint.setAntiAliAndroid Studio(true);
    paint.setTextSize(25);
    paint.setColor(Color.WHITE);
    Paint paint2=new Paint();
    paint2.setStyle(Paint.Style.STROKE);
    paint2.setAntiAliAndroid Studio(true);
    paint2.setColor(Color.RED);
    canvAndroid Studio.drawLine(XPoint, YPoint-YLength, XPoint, YPoint, paint);
```

- 不同的模块有不同的曲线要求,所以根据模块来绘制。实现代码如下:

```
if (swi!=8) {
for (int i=0; i*YScale < YLength; i++) {
    if (i==0 ||i==10 ||i==20 ||i==30 ||i==40 ||i==50
        ||i==60 ||i==70 ||i==80 ||i==90 ||i==100
        ||i==110 ||i==120 ||i==130 ||i==140) {
        canvAndroid Studio.drawLine(XPoint, YPoint-i*YScale, XPoint+5, YPoint
            -i*YScale, paint);
        canvAndroid Studio.drawText(YLabel[i], XPoint-50, YPoint-i*YScale,
            paint);
    } else {
    if (swi==0) {
    if (i==75) {
        canvAndroid Studio.drawLine(XPoint, YPoint-i*YScale, XPoint+XLength,
            YPoint-i*YScale, paint);
canvAndroid Studio.drawText(getResources().getString(R.string.dpm_75),
    XPoint+XLength-20, YPoint-i*YScale-15, paint);
}
} else if (swi==4) {
    if (i==57) {
        canvAndroid Studio.drawLine(XPoint, YPoint-i*YScale, XPoint+XLength,
            YPoint-i*YScale, paint);
        canvAndroid Studio.drawText(getResources().getString(R.string.dpm_37),
            XPoint+XLength-20, YPoint-i*YScale-15, paint);
        }
    }
    }
}
    } else {
YScale=4;
for (int i=0; i*YScale<YLength; i++) {          //判断这几个数值是否有效
    if (i==0 ||i==10 ||i==20 ||i==30 ||i==40 ||i==50
        ||i==60 ||i==70 ||i==80 ||i==90 ||i==100
) {
    canvAndroid Studio.drawLine(XPoint, YPoint-i*YScale, XPoint+5, YPoint
        -i*YScale, paint);
    canvAndroid Studio.drawText(YLabel[i], XPoint-50, YPoint-i*YScale,
    paint);
    }
```

可穿戴设备模块综合设计 第 5 章

```
        }
    }
canvAndroid Studio.drawText(getResources().getString(R.string.company)+company,
        XPoint-30, YPoint-YLength-20, paint);
canvAndroid Studio.drawLine(XPoint, YPoint-YLength, XPoint+XLength, YPoint
        -YLength, paint2);
canvAndroid Studio.drawLine(XPoint, YPoint, XPoint+XLength, YPoint, paint);
```

● 绘制曲线。实现代码如下：

```
if (data.size()>1){
    if (swi==4){
    int xy=60;
for (int i=1; i<data.size(); i++) {      //画笔画线条
    canvAndroid Studio.drawLine((float) (XPoint+(i-1)*XScale),
        (float) (YPoint-data.get(i-1)*YScale)-xy,
        (float) (XPoint+i*XScale),
        (float) (YPoint-data.get(i)*YScale)-xy, paint);
    }
    } else {
    for (int i=1; i<data.size(); i++) {    //画笔画线条
    canvAndroid Studio.drawLine(XPoint+(i-1)*XScale,
        YPoint-data.get(i-1)*YScale, XPoint+i*XScale,
        YPoint-data.get(i)*YScale, paint);
    }
  }
}
```

● 因为曲线需要动态显示，所以需要不断地更新数据调用 invalidate( )来重写 OnDraw( )函数。实现代码如下：

```
private static clAndroid Studios HearHandler extends Handler{
    private WeakReference<RoundView> mActivityReference;
    HearHandler(RoundView activity){
        mActivityReference=new WeakReference<>(activity);
    }              //进去一个新的 Activity
    @Override
    public void handleMessage(Message msg){
        final RoundView activity = mActivityReference.get();
```

```
            if (activity!=null) {
                if (msg.what==1){
                    if (activity.data.size()>=activity.MaxDatAndroid Studioize){
                        activity.data.remove(0);    //删除数据
                    }
                    activity.invalidate();
                }
            }
        }
    }
}
```

- 在布局 fragment_hear.xml 中声明自定义的 View。实现代码如下：

```
<activity.lsen.wearabledevice.view.RoundView
    android:id="@+id/hear_roundview"
    android:layout_width="match_parent"
    android:layout_height="0dp"
    android:layout_gravity="center"
    android:layout_marginTop="10dp"
    android:layout_weight="2" />
```

运行后的效果如图 5-29 所示。

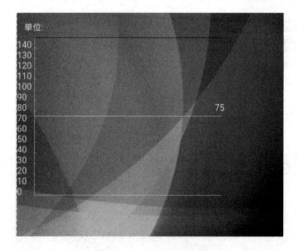

图 5-29　效果图

3. HeartFragment 中使用 RoundProgressBar 和 RoundView

- 在 onCreateView 中使用视图。实现代码如下：

```
public View onCreateView(LayoutInflater inflater, @Nullable ViewGroup container, @Nullable Bundle savedInstanceState){
```

```
View view=inflater.inflate(R.layout.fragment_hear, container, false);
unbinder=ButterKnife.bind(this, view);
hearRoundview.swi=0;          //初始化数值
hearRoundview.data=pullData;
return view;
}
```

● 把收到的心率数据设置到视图中。实现代码如下:

```
private static clAndroid Studios HearHandler extends Handler {
    private WeakReference<HeartFragment> mActivityReference;
    HearHandler(HeartFragment activity) {
        mActivityReference=new WeakReference<>(activity);
    }
    @Override
    public void handleMessage(Message msg) {
        final HeartFragment activity=mActivityReference.get();
        if (activity!=null){
            switch (msg.what) {
                cAndroid Studioe 1:
                    int value=(Integer) msg.obj;
                    if (activity.oldHear!=value){
                        activity.hearRoundbar.setCompany(activity.getResources().getString(R.string.
                            dpm));
                        activity.hearRoundbar.setProgress(value);    //初始化数值
                        activity.hearRoundview.swi=0;
                        activity.pullData.add(value);
                        if(activity.pullData.size()>activity.hearRoundview.MaxDatAndroid Studioize){
                            activity.pullData.remove(0);             //删除数据
                        }
                        activity.hearRoundview.data=activity.pullData;
                        activity.hearRoundview.company=activity.getResources().getString(R.string.dpm);
                        activity.hearRoundview.handler.sendEmptyMessage(1);
                        activity.oldHear=value;
                    }
                    break;                                            //退出程序
                default:
                    break;
            }
        }
    }
}
```

5.4 手腕佩戴式计步模块

5.4.1 运动传感器原理及发展历程

目前,人们普遍认为是 1850 年法国的物理学家莱昂·傅科为了研究地球自转,发明了陀螺仪。那个时代的陀螺仪可以理解成把一个高速旋转的陀螺放到一个万向支架上面。因为陀螺的旋转轴在高速旋转时保持稳定,人们就可以通过陀螺的方向来辨认方向,确定姿态,计算角速度。

万向支架可以保证无论怎么转动,陀螺都不会倒。万向支架最早可以追溯到中国几千年前的香炉。如图 5-30 所示为最早的陀螺仪。

图 5-30 最早的陀螺仪

最早的陀螺仪都是机械式的,里面有高速旋转的陀螺,而机械对加工精度有很高的要求,还会受震动影响,因此以机械陀螺仪为基础的导航系统精度一直都不太高。于是,人们开始寻找更好的办法,利用物理学上的进步,发展出激光陀螺仪、光纤陀螺仪,以及微机电陀螺仪(MEMS)。现代的光纤陀螺仪如图 5-31 所示。

图 5-31 现代的光纤陀螺仪

现代的陀螺仪的原理和传统的机械陀螺仪已经完全不同。光纤陀螺仪利用的是萨格纳克(Sagnac)效应,通过光传播的特性,测量光程差计算出旋转的角速度,起到陀螺仪的作用,替代陀螺仪的功能。

微机电陀螺仪如图 5-32 所示,它是利用物理学的科里奥利力,在内部产生微小的电容变化,然后测量电容,计算出角速度,替代陀螺仪。iPhone 和智能手机中所用的陀螺仪,就是微机电陀螺仪(MEMS)。

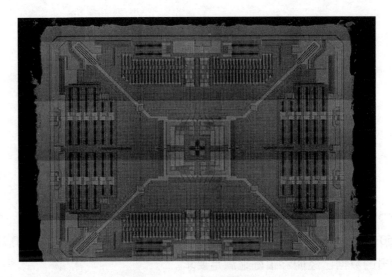

图 5-32 微机电陀螺仪

5.4.2 MPU 6050 传感器的使用

1. MPU 6050 传感器简介

MPU 6050 为全球首例整合性 6 轴运动处理组件,相较于多组件方案,免除了组合陀螺仪与加速器时间轴之差的问题,减少了大量的封装空间。

该传感器以数字输出 6 轴或 9 轴的旋转矩阵、四元数(quaternion)、欧拉角格式(Euler Angle forma)的融合演算数据,具有 131 LSBs 敏感度与全格感测范围为 ±250、±500、±1000 与 ±2000°/s 的 3 轴角速度感测器(陀螺仪)。图 5-33 所示为 MPU 6050 的电路图。

2. MPU 6050 数据采集电路

(1) MPU 6050 传感器坐标系定义:令芯片表面朝向自己,将其表面文字转至正确角度,此时,以芯片内部中心为原点,水平向右的为 X 轴,竖直向上的为 Y 轴,指向自己的为 Z 轴。其示意图如图 5-34所示。

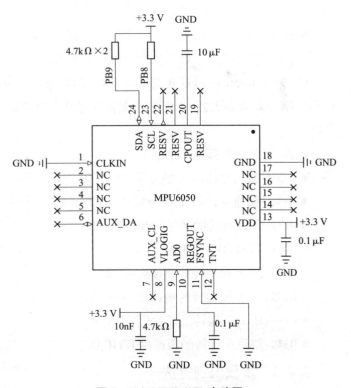

图 5-33 MPU 6050 电路图

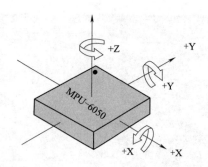

图 5-34　MPU 6050 传感器坐标系

(2) MPU 6050 传感器读/写操作：

● MPU 6050 芯片内自带了一个数据处理子模块 DMP，已经内置了滤波算法，在许多应用中使用 DMP 输出的数据已经能够很好地满足要求。

● MPU 6050 的数据写入和读出均通过其芯片内部的寄存器实现，这些寄存器的地址都是 1 个字节，也就是 8 位的寻址空间。在每次向器件写入数据前要先指定器件的总线地址，MPU 6050 的总线地址是 0x68(AD0 引脚为高电平时地址为 0x69)。然后，写入一个字节的寄存器起始地址，再写入任意长度的数据。这些数据将被连续地写入到指定的起始地址中，超过当前寄存器长度的将写入到后面地址的寄存器中。

● 读出和写入一样，要先写一个字节的寄存器起始地址，接下来将指定地址的数据读到缓存中，并关闭传输模式。最后从缓存中读取数据。

(3) MPU 6050 传感器数据采集：写入的数据位于 0x3B~0x48 这 14 个字节的寄存器中。这些数据会被动态更新，更新频率最高可达 1 000 Hz。下面列出相关寄存器的地址、数据的名称。

注意：每个数据都是 2 个字节。

● 0x3B：加速度计的 X 轴分量 ACC_X。

● 0x3D：加速度计的 Y 轴分量 ACC_Y。

● 0x3F：加速度计的 Z 轴分量 ACC_Z。

● 0x41：当前温度 TEMP。

● 0x43：绕 X 轴旋转的角速度 GYR_X。

● 0x45：绕 Y 轴旋转的角速度 GYR_Y。

● 0x47：绕 Z 轴旋转的角速度 GYR_Z。

● 0x48：绕原点旋转的角速度 GYR_O。

5.4.3　步伐识别算法

1. MPU 6050 传感器计步原理

用户在水平步行运动中，垂直和前进两个加速度会呈现周期性变化。在步行收脚的动作中,由于重心向上单只脚触地,垂直方向加速度是呈正向增加的趋势,之后继续向前,重心下移两脚触底,加速度相反。水平加速度在收脚时减小,在迈步时增加。图 5-35 所示为人体步行示意图。

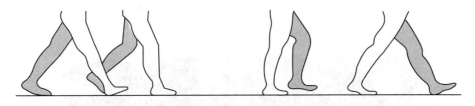

图 5-35　人体步行示意图

2. MPU 6050 传感器计步原理

通过采集数据图表可以看到,在步行运动中,垂直和前进产生的加速度与时间大致为一个正弦曲线,而且在某点有一个峰值。其中,垂直方向的加速度变化最大,通过对轨迹的峰值进行检测计算和加速度阈值决策,即可判断用户是否处于步行状态,实时计算用户运动的步数,还可依此进一步估算用户步行距离。其采集数据图表如图 5-36 所示。

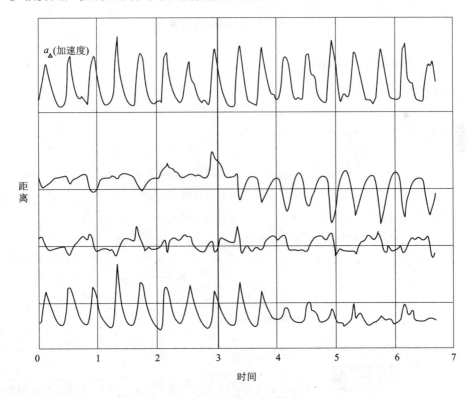

图 5-36　采集数据图表

合理计算步数需要考虑到 3 个方面:

(1)要综合计算 3 个方向的加速度的矢量长度变化,记录步行轨迹。

(2)峰值检测,通过矢量长度的变化,可以判断目前加速度的方向,并和上一次保存的加速度方向进行比较。如果是相反的,即是刚过峰值状态,则进入计步逻辑进行计步,否则舍弃。通过对峰值的次数累加,可得到用户步行的步伐。

(3)去除干扰,判断峰值时通过设置阈值和步频去除干扰,使得计步更准确。

图 5-37 所示为 MPU 6050 传感器图。

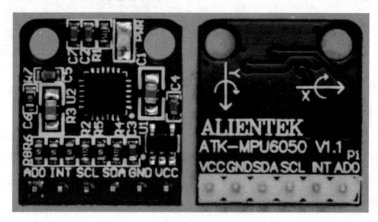

图 5-37　MPU 6050 传感器图

5.4.4　运动传感代码解析

(1)读取加速度数据。实现代码如下:

```
void motionDeviceHandler(void)
{
    static uint8_t timeCnt = 255;     //周期采样计数变量,初始暂缓 255 ms 执行
    struct motionData_t acc_date;     //采样存储结构体
    if (0 == timeCnt) {                //周期采样判断
        timeCnt = MOTION_READ_FREP;   //周期采样赋值50ms
        acc_date.acc_x = GetData(&motionIIC, ACCEL_XOUT_H);  //IIC 接口将 X 轴数值读出
        acc_date.acc_y = GetData(&motionIIC, ACCEL_YOUT_H);  //IIC 接口将 Y 轴数值读出
        acc_date.acc_z = GetData(&motionIIC, ACCEL_ZOUT_H);  //IIC 接口将 Z 轴数值读出
        acc_date.gyro_x = GetData(&motionIIC, GYRO_XOUT_H);  //陀螺仪 X 轴读出
        acc_date.gyro_y = GetData(&motionIIC, GYRO_YOUT_H);  //陀螺仪 Y 轴读出
        acc_date.gyro_z = GetData(&motionIIC, GYRO_ZOUT_H);  //陀螺仪 Z 轴读出
        acc_date.time = getSystemMsTime();
```

```
            motionRingQInsert( & acc_date, & motionQBuff);        //将采样的数据存入队列中

        } else {
        timeCnt --;
        //周期计数变量自减
        }
    }
```

(2)读出缓冲队列数据。实现代码如下：

```
    bool motionRingQCheckout (struct motionData_t * date, struct motionRingQ_t * queuePtr)
    {
        if (0==date ||0==queuePtr){
            return FALSE;
        }
        if (queuePtr->inIndex ==queuePtr->outIndex){
        return FALSE;
        }
        queuePtr->outIndex ++;
        queuePtr->outIndex %=queuePtr->qSize;
        * date =queuePtr->qbuff[queuePtr->outIndex];
        return TRUE;
    }
```

motionRingQCheckout( )函数是一个环形队列函数，入参有两个，分别是存储加速度原始数据的 date 和用于保存环形队列数据的结构体指针 * queuePtr。函数返回则是环形队列是否已空，没满返回 TRUE、已空则返回 FALSE。

环形队列通过对插出计数变量 queuePtr->outIndex 进行% 运算来判断队列是否首尾相连。如果是，则重置 queuePtr->outIndex 为零。

(3)计步数值存入 IIC 缓冲区。实现代码如下：

```
    uint8_t motion_getData(uint8_t* buff)
    {
        buff[0] = (stepCount >>24) & 0xFF;
        buff[1] = (stepCount >>16) & 0xFF;
        buff[2] = (stepCount >>8) & 0xFF;
        buff[3] = (stepCount >>0) & 0xFF;
```

```
buff[4] = m_acc_date.acc_x >> 8;
buff[5] = m_acc_date.acc_x;
buff[6] = m_acc_date.acc_y >> 8;
buff[7] = m_acc_date.acc_y;
buff[8] = m_acc_date.acc_z >> 8;
buff[9] = m_acc_date.acc_z;
```

motion_getData( )函数在 iic_irqHandler( )函数中调用,其中 motion_getData( )函数返回的是自己的数据长度,数组指针返回的则是要传输的步伐数据和六轴数据。

只有在主 IIC 器件询问数据时才会触发数据的发送,所以整个过程是被动的。实现代码如下:

```
buff[10] = m_acc_date.gyro_x >> 8;
buff[11] = m_acc_date.gyro_x;
buff[12] = m_acc_date.gyro_y >> 8;
buff[13] = m_acc_date.gyro_y;
buff[14] = m_acc_date.gyro_z >> 8;
buff[15] = m_acc_date.gyro_z;
return 16;
}
```

5.4.5 手机 APP 软件的开发和功能

1. 加速度模块协议体具体定义(见表 5 - 4)

表 5 - 4 加速度模块协议体具体定义

设 备 号	有效数据长度	有效数据区						
		计步值	ACC_X	ACC_Y	ACC_Z	GYR_X	GYR_Y	GYR_Z
0x03	0x10	4 BYTE	2 BYTE	2 BYTE	2 BYTE	2 BYTE	2 BYTE	2 BYTE

其中 ACC 为重力加速度,换算方式:ACC_X/16384 = 重力加速度_x;GYR 为陀螺仪,换算方式,GYR _X/16.40 = 角速度_x。

数据解析:第 4 章讲到通过 onCharacteristicChanged( )函数回调拿到可穿戴模块的数据 20 个 byte 的数组。通过判断 byte 数组的第一个字节可以知道当前返回的模块设备号;判断 byte 数组的第二个字节可以知道当前数组的有效字节长度;判断 byte 数组的第 3 ~ 6 个字节可以知道加速度的计步整数数值;判断 byte 数组的第 7 ~ 8 个字节可以知道加速度的 ACC_X 数值;依此类推,可得到 ACC_Y、ACC_Z、GYR_X、GYR_Y、GYR_Z 的数值。实现代码如下:

```
cAndroid Studioe 0x03:
int[] valueInt = new int[7];                            // 判断 byte 数组
    int int_0 = value[2] >= 0? value[2]:value[2]+256;
    int int_2 = value[4] >= 0? value[4]:value[4]+256;
    int int_3 = value[5] >= 0? value[5]:value[5]+256;
    valueInt[1] = checkInt(value[6], value[7]);         // 判断 byte 数组
    valueInt[2] = checkInt(value[8], value[9]);
    valueInt[3] = checkInt(value[10], value[11]);
    valueInt[4] = checkInt(value[12], value[13]);
    valueInt[5] = checkInt(value[14], value[15]);
    valueInt[6] = checkInt(value[16], value[17]);
```

因为 byte 的一个字节是 8 位,范围为 $-128\sim127$,即 $-2^7\sim2^7-1$,所以在判断数值时,需要判断数值是否大于零。如果小于零,则需要加上 256 将其转换为正整数。

2. 自定义视图 CustomCurveChart

自定义 CustomCurveChart 显示在 WaveFormFragment 中。

(1) 重绘 OnDraw。具体实现代码如下:

```
@Override
protected void onDraw(CanvAndroid Studio canvAndroid Studio){
    super.onDraw(canvAndroid Studio);
    canvAndroid Studio.drawColor(ContextCompat.getColor(getContext(),
        R.color.color1));
    init();                          //重绘 OnDraw
    drawAxesLine(canvAndroid Studio, paintAxes);
    drawCoordinate(canvAndroid Studio, paintCoordinate);
    for (int i = 0; i < dataList.size(); i++){
        drawCurve(canvAndroid Studio, paintCurve, dataList.get(i),
            colorList.get(i));
    }
}
```

(2) 初始化数据值和画笔,初始化六轴画笔。实现代码如下:

```java
private void init(){
    xPoint = margin + marginX;
    yPoint = this.getHeight() - margin; xScale = (this.getWidth() - 2 * margin -
        marginX)/(xLabel);
    yScale = (this.getHeight() - 2 * margin)/(yLabel.length - 1);
    yValue = yScale * xy;                    //初始化数据值
    paintAxes = new Paint();
    paintAxes.setStyle(Paint.Style.STROKE);
    paintAxes.setAntiAliAndroid Studio(true);
    paintAxes.setDither(true);
    paintAxes. setColor (ContextCompat. getColor (getContext ( ), R. color.
        color14));
    paintAxes.setStrokeWidth(4);
    paintCoordinate = new Paint();           //初始化六轴画笔
    paintCoordinate.setStyle(Paint.Style.STROKE);
    paintCoordinate.setDither(true);
    paintCoordinate.setAntiAliAndroid Studio(true);
    paintCoordinate.setColor (ContextCompat. getColor (getContext ( ), R. color.
        color14));
    paintCoordinate.setTextSize(15);
    paintTable = new Paint();                //初始化六轴画笔
    paintTable.setStyle(Paint.Style.STROKE);
    paintTable.setAntiAliAndroid Studio(true);
    paintTable.setDither(true);
    paintTable. setColor (ContextCompat. getColor (getContext ( ), R. color.
        color4));
    paintTable.setStrokeWidth(2);
    paintCurve = new Paint();                //初始化六轴画笔
    paintCurve.setStyle(Paint.Style.STROKE);
    paintCurve.setDither(true);
    paintCurve.setAntiAliAndroid Studio(true);
    paintCurve.setStrokeWidth(3);
    PathEffect pathEffect = new CornerPathEffect(25);
    paintCurve.setPathEffect(pathEffect);
    paintRectF = new Paint();                //初始化六轴画笔
    paintRectF.setStyle(Paint.Style.FILL);
```

```
paintRectF.setDither(true);
paintRectF.setAntiAliAndroid Studio(true);
paintRectF.setStrokeWidth(3);
paintValue = new Paint();            //初始化六轴画笔
paintValue.setStyle(Paint.Style.STROKE);
paintValue.setAntiAliAndroid Studio(true);
paintValue.setDither(true);
paintValue. setColor (ContextCompat. getColor (getContext ( ), R. color.
color1));
paintValue.setTextAlign(Paint.Align.CENTER);
paintValue.setTextSize(15);
```

(3)设置画笔颜色。实现代码如下：

```
colorList.add(R.color.color14);
colorList.add(R.color.color13);
colorList.add(R.color.color25);
colorList.add(R.color.color5);
colorList.add(R.color.color2);
colorList.add(R.color.text_color);
```

(4)绘制 XY 坐标轴。实现代码如下：

```
private void drawAxesLine(CanvAndroid Studio canvAndroid
Studio, Paint paint){
    // X
    canvAndroid Studio.drawLine(xPoint, yPoint, this.getWidth()-margin/6,yPoint,
        paint);
    canvAndroid Studio.drawLine(this.getWidth()-margin/6, yPoint, this.getWidth
        ()-margin/2,yPoint-margin/3, paint);
    canvAndroid Studio.drawLine(this.getWidth()-margin/6,yPoint, this.getWidth
        ()-margin/2, yPoint+margin/3, paint);
    // Y
    canvAndroid Studio.drawLine(xPoint, yPoint, xPoint, margin / 6, paint);
    canvAndroid Studio.drawLine(xPoint, margin/6, xPoint-margin/3, margin/2,
        paint);
    canvAndroid Studio.drawLine(xPoint, margin/6, xPoint+margin/3, margin/2,
        paint);
}
```

(5)绘制刻度。实现代码如下:

```java
private void drawCoordinate(CanvAndroid Studio canvAndroid Studio, Paint paint){
    // Y 轴坐标
    for (int i=0; i<=(yLabel.length-1); i++){
        paint.setTextAlign(Paint.Align.LEFT);
        int startY = yPoint - i* yScale;
        int offsetX = 0;
        int offsetY;
        offsetY = margin/5;
        //x 默认是字符串的左边在屏幕的位置,y 默认是字符串是字符串的bAndroid
        //Studioeline 在屏幕上的位置
        canvAndroid Studio.drawText(yLabel[i], margin/4 + offsetX,
            startY + offsetY, paint);
    }
}
```

(6)绘制数据。实现代码如下:

```java
private void drawCurve(CanvAndroid Studio canvAndroid Studio, Paint paint, List
        < Integer > data, int color){
    paint.setColor(ContextCompat.getColor(getContext(), color));
    Path path = new Path();
    for (inti=0; i < (data.size()); i++){
        if(i==0){
            path.moveTo(xPoint, toY(data.get(0)));
        } else {
            path.lineTo(xPoint + i* xScale, toY(data.get(i)));
        }

        if(i==data.size() -1){
            path.lineTo(xPoint + i* xScale, toY(data.get(i)));
        }
    }
    canvAndroid Studio.drawPath(path, paint);
}
```

（7）由于坐标的数据较大，需要把数据转化为相应坐标。实现代码如下：

```
private float toY(int num){
    float y;
    try {
        float a = (float) num/5000.0f;
        y = yPoint - a* yScale - 420;
    } catch (Exception e){
        return 0;
    }
    return y;
}
```

（8）不断刷新曲线。实现代码如下：

```
private static clAndroid Studios HearHandler extends Handler {private Weak
   Reference < WaveFormFragment > mActivityReference; HearHandler (WaveFormFragment
   activity){
        mActivityReference = new WeakReference < > (activity);
   }
   @Override
   public void handleMessage(Message msg){
        final WaveFormFragment activity = mActivityReference.get();
        if (activity! = null){
```

（9）在 fragment_motiontemp 中编写布局文件。实现代码如下：

```
<activity.lsen.wearabledevice.view.CustomCurveChartHuman
    android:id = "@ + id/customCurveChartMotion"
    android:layout_width = "match_parent"
    android:layout_height = "match_parent"
    android:gravity = "center"
    android:orientation = "vertical"
    android:paddingBottom = "10dp"
    android:paddingTop = "10dp"/ >
```

(10)给 WaveFormFragment 设置布局。实现代码如下：

```
@Override
public View onCreateView(LayoutInflater inflater, @Nullable ViewGroup container,
    @Nullable Bundle savedInstanceState){
  View view = inflater.inflate(R.layout.fragment_waveform,container, false);
   unbinder = ButterKnife.bind(this, view);
    return view;
}
```

运行后的效果如图 5-38 所示。

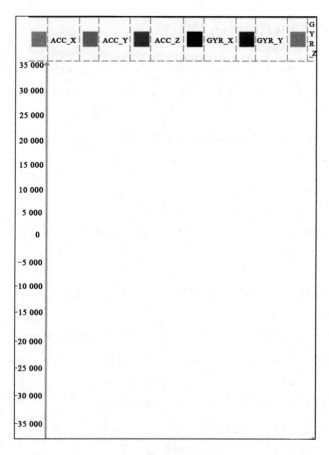

图 5-38　布局效果图

(11)设置自定义视图数据。实现代码如下：

```
activity.data = new ArrayList < > ();          //定义数组成员信息
activity.data.add(activity.data1);
activity.data.add(activity.data2);
activity.data.add(activity.data3);
activity.data.add(activity.data4);
activity.data.add(activity.data5);
activity.data.add(activity.data6);
activity.customCurveChart1.dataList = new ArrayList < > ();
                                               //定义数组成员信息
activity.customCurveChart1.dataList = activity.data;
activity.customCurveChart1.xLabel = activity.data1.size();
activity.customCurveChart1.handler.sendEmptyMessage(1);
break;
```

(12)显示当前加速度的实时步数。实现代码如下:

```
public void handleMessage(Message msg) {
    final WaveFormFragment activity = mActivityReference.get();
    if (activity ! = null) {
    switch (msg.what) {
        cAndroid Studioe 1:
        int[] value = (int[]) msg.obj;              //显示当前加速度
        activity.waveNumberTv.setText("步数:".concat(String.valueOf(value[0])));
        activity.data1.add(value[1]);
        activity.data2.add(value[2]);
        activity.data3.add(value[3]);
        activity.data4.add(value[4]);
        activity.data5.add(value[5]);
        activity.data6.add(value[6]);
    }}
}
```

5.5 腕部触感提醒模块

5.5.1 微型振动马达原理及使用注意事项

主要用于手机的微型振动马达属于直流有刷电机,马达轴上面有一个偏心轮,当马达转动时,偏心轮的圆心质点不在电机的转心上,使得马达处于不断的失去平衡状态,由于惯性作用引起震动。微型振动马达的分解图如图5-39所示。

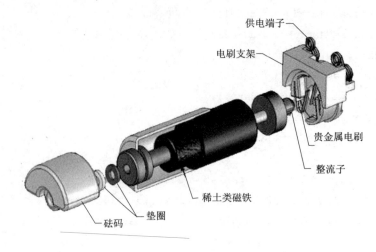

图5-39 微型振动马达的分解图

使用马达的注意事项如下：

（1）马达在其标称的额定电压下工作时综合性能优良，建议手机电路设计时工作电压尽量接近额定电压设计。

（2）给马达供电的控制模块应考虑其输出阻抗尽量小，防止负载时输出电压大幅度下降，影响振感。

（3）柱式电机检验或测试堵转电流时，堵转时间不宜过长（小于5 s为宜），因为堵转时所有的输入功率都转化为热能（$P = I^2 R$），时间过长可能导致线圈温升偏高而变形，影响性能。

（4）带安装支架的马达在设计定位卡槽时，与外壳的间隙不能太大，否则有可能产生附加振动（机械噪声），采用橡胶套固定可有效避免机械噪声，但应注意外壳上定位槽与橡胶套应采用过盈配合，否则会影响马达的振动输出，振感下降。

（5）中转或使用时避免靠近强磁区，否则有可能使马达磁钢表磁扭曲而影响性能。

5.5.2 微型振动马达电路解析

1. 微型振动马达参数

微型马达的实物图如图5-40所示。

本次采用的马达是直流3 V的小型振动马达，主要应用在DIY航模、玩具等。

设计马达驱动电路要注意电气参数的极限范围，如额定电压、堵转电流等。电气参数直接关联到电路驱动应该怎么设计。

图5-40 微型马达的实物图

2. 电机驱动原理

电机的起/停控制，最简单、最原始的方法是在电机与电源之间加一个机械开关,或者用继电器的触点控制。

现在比较流行的方法是用开关晶体管来代替机械开关,无触点、无火花干扰,速度快。当输入端为低电平时,电机无电流而处于停止状态。如果输入端为高电平时,电机中有电流,因此电机起动运转。电阻保护二极管,防止反电动势损坏晶体管。消除射频干扰而外加的 1 kΩ 基极限流电阻,限制 TIP41 的基极电流。在 6 V 电源时,基极电流不超过 300 mA。在这种情况下,提供电机的最大电流为 1 A 左右。电路原理图如图 5 – 41 所示。

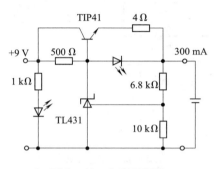

图 5 – 41　电路原理图

本项目的电路设计是通过 NPN 晶体管器件对马达进行开关控制,当 PC4 端口输出高电平时马达电源接通马达振动,当输出低电平时电源断开马达停止振动。具体实现电路如图 5 – 42 所示。

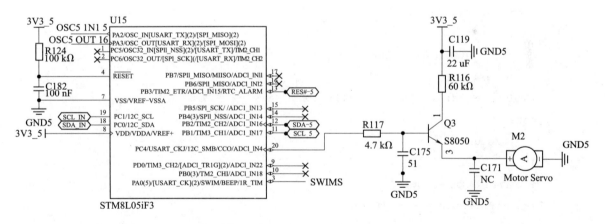

图 5 – 42　电路图

5.5.3　振动提示程序解析

振动提醒模块是通过接收蓝牙模块 IIC 总线发送的指令进行工作的,所以默认情况振动提醒模块是不会振动的。

整体运行分为两个分支:接收参数更改的分支模块和执行设置参数的马达控制模块,这两个分支并行存在。其软件流程图如图 5 – 43 所示。

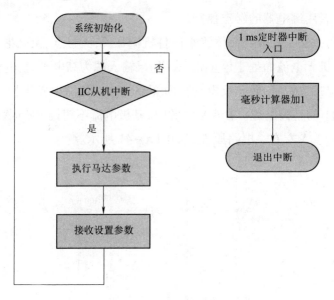

图 5-43 软件流程图

接收设置参数代码如下:

```
void motorSetWokeInfo(enum motro_workMode_t mode, uint8_t newRunTime){
    switch (mode){                              //设置接收模式
    cAndroid Studioe MOTOR_WORK_MODE_HOLD:
    cAndroid Studioe MOTOR_WORK_MODE_SHORT:
        motorCtrlOut(OUT_HIGH);
        break;
    cAndroid Studioe MOTOR_WORK_MODE_IDLE:      //接收设置参数
    default:
        return;
    }
    motorWorkMode = mode;
    runingTimeCount = newRunTime;
}
```

其中,motorSetWokeInfo()函数负责设置振动模式和振动时间。

motorSetWokeInfo()在 motor_recvDataProcess()函数中被调用,而 motor_recvDataProcess()函数则在 iic_irqHandler()函数中被调用,所以 motorSetWokeInfo()函数是运行在中断服务函数中的,在修改函数时就不要在里面添加延时了。

执行马达参数代码如下:

```
switch (motorWorkMode){
cAndroid Studioe MOTOR_WORK_MODE_HOLD:
    if (0 == runingTimeCount){
        motorWorkMode = MOTOR_WORK_MODE_IDLE;
        motorRuning(FALSE);                       //让马达停止运转
        motorDisplay(motorWorkMode, 0);
    } else {
        motorRuning(TRUE);
        runingTimeCount --;                        //减少运转时间
        motorDisplay(MOTOR_WORK_MODE_HOLD, runingTimeCount);
    }
    break;
cAndroid Studioe MOTOR_WORK_MODE_SHORT:
    if (shortRunTimeCnt == 0){
        if (0 < runingTimeCount){                 //判断马达运转状态
            if (RESET == motorCtrlIn()){
                motorCtrlOut(OUT_HIGH);
            } else {
                motorCtrlOut(OUT_LOW);
                runingTimeCount --;
                motorDisplay(MOTOR_WORK_MODE_SHORT, runingTimeCount);
            }
            shortRunTimeCnt = SHORT_RUN_TIME_DEF;
        } else {
            motorWorkMode = MOTOR_WORK_MODE_IDLE;
            motorDisplay(motorWorkMode, 0);//显示列表
        }
    } else {
        shortRunTimeCnt --;
    }
}
break;
```

由于篇幅原因删减掉了部分代码。

执行函数是以 switch 的形式执行电机状态的切换以及振动时间的倒计时。

motorDeviceHandler( )函数是建立在定时器中断基础上运行的,定时器中断一次函数运行一次,所以在长时间的马达定时振动下程序的主线程是不会被长时间占用的,这样程序就能迅

速响应新下发的指令。

5.5.4 手机 APP 软件的开发和功能

1. 马达模块协议体具体定义(见表 5-5)

表 5-5 马达模块协议体具体定义

设备号	长度	有效数据区	
		振动方式	振动参数
0x02	0x02	0:连续振动 1:短促振动	连续振动:振动单位取值 0~255(每个振动单位约 10ms) 短促振动:短促振动的次数 0~20

数据发送,实现代码如下:

```
if (!isSendData){
    byte[] value = new byte[20];
        value[0] = 0x02;
        value[1] = 0x02;
if (!isCheckBox){
        value[2] = 0x00;
        value[3] = (byte) number;
        interval = number* 100;
} else {
        value[2] = 0x01;
        value[3] = (byte) (number +1);
        interval = number* 1000;
}
```

当选择连续振动时,value[2]写入数值 0x00;选择短促振动时,value[2]写入数值 0x01,以此来区分当前发送的指令是哪种振动模块。

2. 自定义视图

(1)定义 MotorFragment 的视图 fragment_motor。实现代码如下:

```
<? xml version = "1.0" encoding = "utf -8"? >
<LinearLayout xmlns:android = "http://schemAndroid Studio.android.com/apk/res/android"
    android:layout_width = "match_parent"
    android:layout_height = "match_parent"
    android:background = "@mipmap/gai_112"
    android:orientation = "vertical" >
```

```xml
<LinearLayout                              //自定义文本框
    android:layout_width="match_parent"
    android:layout_height="match_parent"
    android:layout_margin="40dp"
    android:background="@drawable/shape_layout"
    android:gravity="center"
    android:orientation="vertical" >
<TextView                                  //自定义文本框
    android:layout_width="match_parent"
    android:layout_height="0dp"
    android:layout_weight="2" />
<TextView                                  //自定义文本框
    android:layout_width="match_parent"
    android:layout_height="0dp"
    android:layout_marginEnd="@dimen/motor_left_right"
    android:layout_marginStart="@dimen/motor_left_right"
    android:layout_marginTop="@dimen/motor_left_right"
    android:layout_weight="1"
    android:gravity="center|start"
    android:text="@string/pattern"
    android:textColor="@color/white"
    android:textSize="@dimen/text_title" />
<LinearLayout                              //自定义布局
    android:layout_width="match_parent"
    android:layout_height="0dp"
    android:layout_marginStart="@dimen/motor_left_right"
    android:layout_marginTop="@dimen/motor_left_right"
    android:layout_weight="1"
    android:orientation="horizontal" >
<TextView                                  //自定义文本框
    android:layout_width="0dp"
    android:layout_weight="3"
    android:layout_height="match_parent"
    android:gravity="center|start"
    android:text="@string/vibration_continuity"
    android:textColor="@color/white"
    android:textSize="@dimen/text_size" />
<CheckBox                                  //自定义选择框
```

```xml
        android:id = "@ + id/continuity_checkbox"
        android:layout_width = "0dp"
        android:layout_height = "match_parent"
        android:layout_weight = "1" />
    </LinearLayout >
    <LinearLayout                           //自定义布局
        android:layout_width = "match_parent"
        android:layout_height = "0dp"
        android:layout_marginStart = "@dimen/motor_left_right"
        android:layout_marginTop = "@dimen/motor_left_right"
        android:layout_weight = "1"
        android:orientation = "horizontal" >
        <TextView                           //自定义文本框
            android:layout_width = "0dp"
            android:layout_height = "match_parent"
            android:layout_weight = "3"
            android:gravity = "center |start"
            android:text = "@string/vibration_short"
            android:textColor = "@color/white"
            android:textSize = "@dimen/text_size" />
        <CheckBox                           //自定义选择框
            android:id = "@ + id/short_checkbox"
            android:layout_width = "0dp"
            android:layout_height = "match_parent"
            android:layout_weight = "1" />
    </LinearLayout >
    <TextView                               //自定义文本框
        android:id = "@ + id/company_tv"
        android:layout_width = "match_parent"
        android:layout_height = "0dp"
        android:layout_marginEnd = "@dimen/motor_left_right"
        android:layout_marginStart = "@dimen/motor_left_right"
        android:layout_marginTop = "@dimen/motor_left_right"
        android:layout_weight = "1"
        android:gravity = "center |start"
        android:text = "@string/company"
        android:textColor = "@color/white"
```

```xml
        android:textSize = "@dimen/text_title" / >
    < SeekBar                     //自定义进度条
        android:id = "@ + id/seekBar"
        android:layout_width = "match_parent"
        android:layout_height = "0dp"
        android:layout_marginEnd = "@dimen/motor_left_right"
        android:layout_marginStart = "@dimen/motor_left_right"
        android:layout_marginTop = "@dimen/motor_left_right"
        android:layout_weight = "1" / >
    < TextView                    //自定义文本框
        android:id = "@ + id/motor_sendTv"
        android:layout_width = "match_parent"
        android:layout_height = "0dp"
        android:layout_marginBottom = "@dimen/motor_left_right"
        android:layout_marginEnd = "@dimen/motor_left_right"
        android:layout_marginStart = "@dimen/motor_left_right"
        android:layout_marginTop = "@dimen/motor_left_right"
        android:layout_weight = "1"
        android:background = "@drawable/button_bar"
        android:gravity = "center"
        android:text = "@string/send_out"
        android:textColor = "@color/white"
        android:textSize = "@dimen/text_size" / >
```

运行后 XML 布局效果如图 5 – 44 所示。

图 5 – 44　效果图

（2）为 MotorFragment 设置视图。实现代码如下：

```java
public View onCreateView(LayoutInflater inflater, @Nullable ViewGroup container,
@Nullable Bundle savedInstanceState){
    View view = inflater.inflate(R.layout.fragment_motor, container, false);
    unbinder = ButterKnife.bind(this, view);
    init();
    return view;
}
```

（3）初始化进度条及模式。实现代码如下：

```java
private void init(){                      //初始化进度条及模式
    continuityCheckbox.setChecked(true);
    shortCheckbox.setChecked(false);
    companyTv.setText(getResources().getString(R.string.company_100));
    seekBar.setMax(25);
    seekBar.setProgress(1);
}
```

（4）当选择模式时，改变进度条的最大值。实现代码如下：

```java
continuityCheckbox.setOnCheckedChangeLisener(new CompoundButton.OnChecked
    ChangeListener(){
@Override
public void onCheckedChanged(CompoundButton buttonView, boolean isChecked){
    if (isChecked){                      //设置最大数值
    seekBar.setMax(25);
    seekBar.setProgress(1);
    isCheckBox = false;
    companyTv.setText(getResources().getString(R.string.company_100));
    shortCheckbox.setChecked(false);
} else {
    shortCheckbox.setChecked(true);
}
```

```
        }
});
shortCheckbox.setOnCheckedChangeListener(new CompoundButton.OnCheckedChange
    Listener(){
    @Override
    public void onCheckedChanged(CompoundButton buttonView, boolean isChecked){
        if (isChecked){
            seekBar.setMax(20);
            seekBar.setProgress(1);
            companyTv.setText(getResources().getString(R.string.company_1));
            isCheckBox = true;
            continuityCheckbox.setChecked(false);
        } else {
            continuityCheckbox.setChecked(true);
        }
    }
});
```

(5) 发送振动数据。实现代码如下：

```
@OnClick(R.id.motor_sendTv)
public void onViewClicked(){
    mainActivity.mainHandler.sendEmptyMessage(7);
    if (mainActivity.isConnectState){
        if (!isSendData){
            byte[] value = new byte[20];
            value[0] = 0x02;
            value[1] = 0x02;
            if (!isCheckBox){          //勾选选择框
                value[2] = 0x00;
                value[3] = (byte) number;
                interval = number * 100;
            } else {
                value[2] = 0x01;
                value[3] = (byte) (number + 1);
```

```
                interval = number* 1000;
            }
            if (mainActivity.weData != null){
                mainActivity.weData.setMotor(String.valueOf(value[3]));
            }
            isSendData = true;            //确定已经发送数据
            mainActivity.sendData = value;
            handler.sendEmptyMessage(2);
            handler.sendEmptyMessage(1);
        } else {
            mainActivity.showToAndroid Studiot("马达模块振动中...");
        }
    } else {
        mainActivity.showToAndroid Studiot(getResources().getString(R.
            string.un_connect));
    }
}
```

因为在可穿戴技术项目中,蓝牙模块在发送数据给APP时,APP端不能同时给蓝牙模块发送数据,所有需要在蓝牙模块发送数据给APP后的30 ms的间隔内发送到马达的振动数据,需要进行判断。

(6)如没有收到蓝牙数据,则在100 ms后发送。实现代码如下:

```
cAndroid Studioe 1:
    activity.mainActivity.isSend = false;    //如没有收到蓝牙数据
    activity.handler.postDelayed(activity.runnable, 100);
    break;
    private Runnable runnable = new Runnable(){
        @Override
        public void run(){
            if (!mainActivity.isSend){       //在100 ms后发送
                mainActivity.writeCharacteristic(mainActivity.sendData);
                mainActivity.isSend = true;
            }
            handler.removeCallbacks(this);
        }
    };
```

(7) 发送马达振动数据后,需要等待马达模块振动结束才能发送下一条振动数据,所以需要时间判断。实现代码如下:

```
cAndroid Studioe 2:
    activity.handler.postDelayed(activity.runnable2, activity.interval);
    break;
```

5.6 紫外线超测量模块

5.6.1 紫外线的检测原理及注意事项

紫外线是指阳光中波长 10~400 nm 的光线,可分为 UVA(波长 320~400 nm,长波)、UVB(波长 290~320 nm,中波)、UVC(波长 100~290 nm,短波)。UVB 致癌性最强,晒红及晒伤作用为 UVA 的 1 000 倍。UVC 可被臭氧层所阻隔。紫外线照射会让皮肤产生大量自由基,导致细胞膜的过氧化反应,使黑色素细胞产生更多的黑色素,并往上分布到表皮角质层,造成黑色斑点。紫外线可以说是造成皮肤皱纹、老化、松弛及黑斑的最大元凶。图 5-45 所示为紫外线对皮肤的危害。

图 5-45 紫外线对皮肤的危害

紫外线能使许多物质激发荧光，很容易让照相底片感光。当紫外线照射人体时，能促使人体合成维生素 D，以防止患佝偻病，经常让小孩晒晒太阳就是这个道理。紫外线还具有杀菌作用，医院里的病房就利用紫外线消毒。但过强的紫外线会伤害人体，应注意防护。玻璃、大气中的氧气和高空中的臭氧层，对紫外线都有很强的吸收作用，能吸收掉太阳光中的大部分紫外线，因此能保护地球上的生物，使它们免受紫外线伤害。图 5-46 所示为地球臭氧层对紫外线的过滤作用。

图 5-46　地球臭氧层对紫外线的过滤作用

1. 紫外线传感器的发展历史

紫外线传感器是利用光敏元件将紫外线信号转换为电信号的传感器，它的工作模式通常分为两类：光伏模式和光导模式。光伏模式是指不需要串联电池，串联电阻中有电流，而传感器相当于一个小电池，输出电压，但是制作比较难，成本比较高；光导模式是指需要串联一个电池工作，传感器相当于一个电阻，电阻值随光的强度变化而变化，这种制作容易，成本较低。

2. 非接触式紫外线传感器构造

最早的紫外线传感器是基于单纯的硅，但是根据美国国家标准与技术研究院的指示，单纯的硅二极管也响应可见光，形成本来不需要的电信号，导致精度不高。

在十几年前，日本日亚公司研发出了 GaN 系的晶体，成为 GaN 系的开拓者，并由此开辟了 GaN 系的市场，也由此产生了 GaN 的紫外线传感器，其精度远远高于单晶硅的精度，成为最常用的紫外线传感器材料。

二六族 ZnS 材料已被研发出来，也应用到了紫外线传感器领域，从研发的角度及性能测试上看，其精度比 GaN 系的传感器提高了近 10^5 倍。在一定程度上，ZnS 系的紫外线传感器将与 GaN 系的平分秋色。

(1) 电气特性：
- 采用氮化镓基材料；
- PIN 型光电二极管；
- 光伏工作模式；
- 对可见光无响应；
- 暗电流低；
- 输出电流与紫外指数呈线性关系；
- 符合欧盟 RoHS 指令，无铅、无镉。

(2) 典型应用
- 测量紫外指数：手机、数码照相机、MP4、PDA、GPS 等携式移动产品。
- 用于紫外检测器：全部紫外线波段的检测器、单 UV–A 波段检测器、紫外线指数检测器、紫外线杀菌灯辐照检测器。图 5–47 所示为紫外线检测器的实物图。

图 5–47 紫外检测器的实物图

注意事项：

(1) 环境紫外线采集模块与 BLE 模块正确连接起来，在运动模块的 DEBUG 口接入转接板并进行调试，与计算机连接。

(2) 按下 BLE 模块的开关接通电源，BLE 模块的蓝色指示灯亮起，并且在 OLED 屏上显示设备名，说明 BLE 模块正常；运动模块正面的红色 LED 亮，说明采集模块已工作。如果 BLE 模块电池电量不足，可以使用电源线在 BLE 模块或 DEBUG 模块的 micro–usb 座上接入 5 V 电源。

(3) 在计算机上启动 IAR 编译环境，然后打开"代码\\Ultraviolet\\PRG\\Project"的 WEARABLE 工程，并编译和下载到环境紫外线采集模块中。

(4) 通过腕带将紫外线采集模块固定在手腕上或其他肢体上，保持模块不从腕带上脱落，也不会压迫手腕为宜。保证设备固定好后确保 BLE 模块与紫外线采集模块连接稳定，保持 BLE 开关接通。

(5) 打开"可穿戴技术平台"应用程序，点击"连接设备"查看可连接的 BLE 设备，找到与 BLE 模块同名(如 UART–E76EE6)的蓝牙设备，点击连接此设备。如果 APP 界面下方出现了 BLE 模块蓝牙名称(如 UART–E76EE6)，则说明设备连接成功。

(6) 点击主界面的"紫外线"图片，进入运动采集界面，在界面中央可以读到当前的步数。

(7) 将变换环境紫外线模块朝向阳光的角度，观察平板计算机上的环境紫外线值变化。

5.6.2 紫外线传感器电路解析

这里使用 GUVA–S12SD 紫外线传感器，可放置在自然环境中检测 UV 强度。

GUVA – S12SD 电器示例如图 5 – 48 所示。

图 5 – 48　GUVA – S12SD 电器示例

GUVA – S12SD 配合一路运算放大器组成紫外线传感器的采集输出电路。MUC 端则需要配置好 ADC 引脚，从而对输出电路电压进行线性测量。GUVA – S12SD 电路图如图 5 – 49 所示。

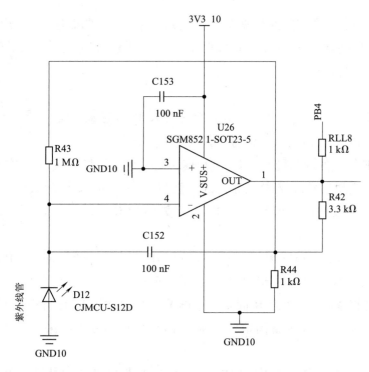

图 5 – 49　GUVA – S12SD 电路图

5.6.3　紫外线传感代码解析

1. MCU 嵌入式系统功能

通过 ADC 采集传感器输出的电压模拟信号，并通过数字滤波算法去除干扰信号，然后将有效的电压信号转化为强度数值通过蓝牙模块传输到 APP 中。

2. 手机 APP 软件功能

收集 MCU 发送来的紫外线强度信号，并将存入到数据库中，形成历史曲线图。

结合 X 轴的时间刻度就可以清楚地了解到,佩戴者一天在户外所受到紫外线的照射强度分别是多少,从而有针对性地做好保护措施。其软件流程图如图 5-50 所示。

在这个过程中,运算放大器将毫伏电压信号放大稳定为 MUC 可读取的电压信号。单片机通过 ADC 采样和固定的周期频率,从而计算出人体的心率。

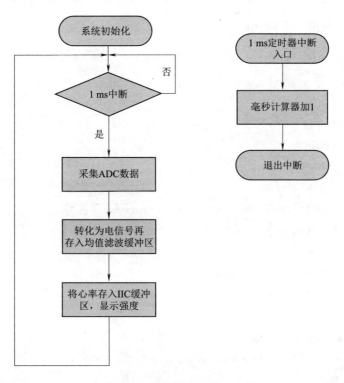

图 5-50 软件流程图

在计算过程中考虑到测量过程有干扰的存在,所以中间使用了均值滤波算法将其中的干扰剔除,最终获得稳定的 UV 强度。

(1) 采集 ADC 数据。实现代码如下:

```
void analogDeviceHandler(void)
{
    static uint8_t timeCnt = 255;          //周期计数变量,初始值255是暂缓启动数值
    struct analogAdcDate_t adcData;        //存储数据结构体
    if (0 == timeCnt){                     //周期采样判断
        timeCnt = ANALOG_READ_FREP;        //周期间隔数值赋予
        adcData.adcM = ADC_GetConversionValue(ADC1);//通过 stm8s 库函数将采样数据
                                           //读出
        anglogDataInput(adcData);          //将读出采样数据存入环形队列中
```

```
        } else {
            timeCnt --;          //周期间隔数值自减
        }
    }
```

(2)存入缓冲区数据。实现代码如下:

```
bool analogRingQInsert(struct analogAdcDate_t date, struct ringQueue_t*
    queuePtr)
{
    int tmpInput;
    if (0 == queuePtr){
        return FALSE;
    }
    tmpInput = queuePtr -> inIndex + 1;      //存入缓冲区代码
    tmpInput % = queuePtr - > qSize;

    if (tmpInput == queuePtr -> outIndex){
        return FALSE;
    }
    queuePtr -> inIndex = tmpInput;
    queuePtr -> qbuff[tmpInput] = date;      //存入缓冲区代码
    return TRUE;
}
```

代码说明:

anglogDataInput()函数嵌套在analogRingQInsert()函数中,所以这里直接讲analogRingQInsert()函数了

analogRingQInsert()函数是一个环形队列函数,入参有两个,分别是存储ADC原始数据的date和用于保存环形队列数据的结构体指针*queuePtr。函数返回则是环形队列是否存储已满,没满返回TRUE,满了则返回FALSE。

环形队列通过对插入计数变量tmpInput进行%运算来判断队列是否首尾相连,如果是,则重置tmpInput为零。

(3)强度转换数据。实现代码如下:

```
void analog_ultravioletProcess(struct analogAdcDate_t data)
{
    uint16_t i;
```

```
uint32_t tmpMeans = 0;
for (i = 1; i<WARE_MEANS_BUFF_SIZE; i ++){
        //将整体缓冲数组的元素整体往左移,空出最右端元素载入新的元素
        wareMeansBuff[i - 1] = wareMeansBuff[i];
        //将本次的缓冲数组进行累加
        tmpMeans += wareMeansBuff[i];
    }
    //将新的计算元素载入数组中
    wareMeansBuff[WARE_MEANS_BUFF_SIZE - 1] = data.adcM;
    //将本次最新的变量进行累加
    tmpMeans + = data.adcM;
    //累加数值均值计算及强度转换
    ultravioletValue = ((tmpMeans* 100)/WARE_MEANS_BUFF_SIZE)/300;
    //强度过域判断
    if (ultravioletValue >100){
       ultravioletValue = 100;
    }
    uint8_t strBuff[17];
    intNumToStr(ultravioletValue,strBuff);
    //格式打印
    strcat(strBuff,"%    ");
    //根据格式打印的内容进行显示
    OLED_ShowString(0, 4, strBuff);
}
```

强度数值存入 IIC 缓冲区代码:

```
uint8_t analog_getData(uint8_t* buff)
{
    buff[0] = ultravioletValue;
    return 1;
}
```

analog_getData( )函数在 iic_irqHandler( ) 函数中调用,其中 analog_getData( )函数返回的是自己的数据长度,数组指针返回的则是要传输的 UV 强度数据。

只有在主 IIC 器件询问数据时才会触发数据的发送,所以整个过程是被动的。

5.6.4 手机 APP 软件的开发和功能

1. 紫外线模块协议体具体定义(见表 5 – 6)

表 5 – 6　紫外线模块协议体具体定义

设 备 号	有效数据长度	有效数据区
0x08	0x01	紫外线辐射强度 0 ~ 100

数据解析:在第 4 章讲到通过 onCharacteristicChanged( ) 函数回调拿到可穿戴模块的数据 20 个 byte 的数组。通过判断 byte 数组的第一个字节可以知道当前返回的模块设备号;判断 byte 数组的第二个字节可以知道当前数组的有效字节长度;判断 byte 数组的第三个字节可以知道紫外线的整数数值。实现代码如下:

```
cAndroid Studioe 0x08:
    if (value[1] ==0x01){
        int va = value[2] > =0 ? value[2] : (value[2] +256);//当前数组的有效字节长度
        if (activity.weData ! = null){
            activity.weData.setUv(String.valueOf(va));
        }
        if (activity.uvFragment ! = null){
            if (activity.currentFragment == activity.uvFragment){
                if (activity.uvFragment.handler ! = null){  //判断 byte 数组的第三个字
                    Message message = new Message();
                    message.what =1;
                    message.obj = va;
                    activity.uvFragment.handler.sendMessage(message);
                }
            }
        }
    }
    break;
```

因为 byte 的一个字节是 8 位,范围为 – 128 ~ 127,即 $-2^7 \sim 2^7 -1$,所以在判断紫外线数值时,需要判断数值是否大于零,如果小于零则需要加上 256 将其转换为正整数。

2. 自定义视图

自定义 RoundProgressBar 和自定义 RoundView 显示在 HeartFragment 中。

自定义 RoundProgressBar 的内容包括:

- 自定义 View 的属性。
- 在 View 的构造方法中获得自定义的属性。
- 重写 onMesure(onMesure 不一定是必需的)。
- 重写 onDraw。

(1)自定义 RoundProgressBar:

- 在 res/values/下创建 attrs.xml；自定义 View 的属性。实现代码如下:

```xml
<?xml version = "1.0" encoding = "UTF-8"?>
<resources>
<declare-styleable name = "RoundProgressBar">      ;//自定义 view 的属性
<attr name = "roundColor" format = "color"/>
<attr name = "roundProgressColor" format = "color"/>
<attr name = "roundWidth" format = "dimension"/>
<attr name = "textColor" format = "color" />
<attr name = "textSize" format = "dimension" />
<attr name = "max" format = "integer"/>             //设置最大长度
<attr name = "textIsDisplayable" format = "boolean"/>
<attr name = "style">
<enum name = "STROKE" value = "0"/>
<enum name = "FILL" value = "1"/>
</attr>
</declare-styleable>
</resources>
```

- 在布局 fragment_uv.xml 中声明自定义 View。实现代码如下:

```xml
<activity.lsen.wearabledevice.view.RoundView
    android:id = "@+id/uv_roundview"              //自定义 View
    android:layout_width = "match_parent"
    android:layout_height = "0dp"
    android:layout_gravity = "center"
    android:layout_marginTop = "10dp"
    android:layout_weight = "2" />
```

- 在 View 的构造方法中获得自定义的属性。实现代码如下:

```java
public RoundProgressBar(Context context){
    this(context, null);
```

```java
}
public RoundProgressBar(Context context, AttributeSet attrs){this(context,
    attrs, 0);                                    //声明RoundPro-gressBar
}
//获得我们自定义的属性
public RoundProgressBar(Context context, AttributeSet attrs, int defStyle){
    super(context, attrs, defStyle);
    paint=new Paint();
    TypedArray mTypedArray=context.obtainStyledAttributes(attrs,
        R.styleable.RoundProgressBar);
    roundColor=mTypedArray.getColor(              //获取颜色
        R.styleable.RoundProgressBar_roundColor, Color.RED);
    roundProgressColor=mTypedArray.getColor(    //获取颜色
      R.styleable.RoundProgressBar_roundProgressColor, Color.GREEN);
    textColor=mTypedArray.getColor(               //获取颜色
        R.styleable.RoundProgressBar_textColor, Color.GREEN);
    textSize=mTypedArray.getDimension(
        R.styleable.RoundProgressBar_textSize, 15);
    roundWidth=mTypedArray.getDimension(
        R.styleable.RoundProgressBar_roundWidth, 40);
    max=mTypedArray.getInteger(R.styleable.RoundProgressBar_max, 140);
    textIsDisplayable=mTypedArray.getBoolean(
        R.styleable.RoundProgressBar_textIsDisplayable, true);
    style=mTypedArray.getInt(R.styleable.RoundProgressBar_style, 0);
    mTypedArray.recycle();
}
```

- 重写 OnDraw。实现代码如下：

```java
@Override
protected void onDraw(CanvAndroid Studio canvAndroid Studio){
    super.onDraw(canvAndroid Studio);
    int centre=getWidth()/2;                      //计算得到中心点
    int radius = (int) (centre - roundWidth/2);
```

```java
        paint.setColor(roundColor);                    //设置进度条颜色
        paint.setStyle(Paint.Style.STROKE);
        paint.setStrokeWidth(roundWidth);              //设置定点位置
        paint.setAntiAliAndroid Studio(true);
        //重写 OnDraw
        canvAndroid Studio.drawCircle(centre, centre, radius, paint);
        paint.setStrokeWidth(0);
        paint.setColor(textColor);
        paint.setTextSize(textSize);
        paint.setTypeface(Typeface.DEFAULT_BOLD);
        double percent =(double) (((float) progress/(float) max)* 140);
        percent = formatDouble(percent);
        float textWidth = paint.meAndroid StudioureText(percent + company);
        if (textIsDisplayable && percent!=0 && style == STROKE){canvAndroid Studio.
            drawText(percent + company, centre - textWidth/2, centre + textSize/2,
                paint);
        } else {
            canvAndroid Studio.drawText(percent + company, centre - textWidth/2, centre +
                textSize/2, paint);
        }
        paint.setStrokeWidth(roundWidth);
        paint.setColor(textColor);
        RectF oval = new RectF(centre - radius, centre - radius, centre + radius, centre +
            radius);
    switch (style){
        cAndroid Studioe STROKE: {
            paint.setStyle(Paint.Style.STROKE);
            canvAndroid Studio.drawArc (oval, 0, (float) (360 * progress/max), false,
                paint);
            break;
    }
        cAndroid Studioe FILL: {
            paint.setStyle(Paint.Style.FILL_AND_STROKE);
            if (progress !=0)
        canvAndroid Studio.drawArc(oval,0,(float) (360* progress /max), true, paint);
```

```
            break;
        }
    }
}
```

运行 RoundProgressBar,效果如图 5-51 所示。

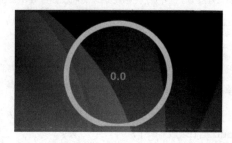

图 5-51　运行 RoundProgressBar 后的效果图

- 在 RoundProgressBar 中设置数值的接口。实现代码如下:

```
public synchronized void setProgress(double progress){
    if(progress>max){
        progress=max;
    }
    if(progress<=max){
        this.progress=progress;
        postInvalidate();
    }
}
```

(2) 自定义 RoundView:

视图未做屏幕适配处理,如果需要可以自己尝试修改。实现代码如下:

```
绘制曲线边线的 XY 轴
protected void onDraw(CanvAndroid Studio canvAndroid Studio){
    super.onDraw(canvAndroid Studio);
    MaxDatAndroid Studioize=XLength/XScale;
    YLabel=new String[YLength/YScale];
    if(swi==0 || swi==8){
        for (int i=0;i<YLabel.length; i++){
```

```
            YLabel[i] = (i) + "";
    }
} else {
    for (int i = 0; i < YLabel.length; i ++){
        YLabel[i] = (i - 20) + "";
    }
}                                    //绘制曲线边线 x
Paint paint = new Paint();
paint.setStyle(Paint.Style.STROKE);
paint.setAntiAliAndroid Studio(true);
paint.setTextSize(25);
paint.setColor(Color.WHITE);
Paint paint2 = new Paint();          //绘制曲线边线 y
paint2.setStyle(Paint.Style.STROKE);
paint2.setAntiAliAndroid Studio(true);
paint2.setColor(Color.RED);
canvAndroid Studio.drawLine(XPoint, YPoint - YLength, XPoint, YPoint, paint);
```

- 不同的模块有不同的曲线要求,所以根据模块来绘制。实现代码如下:

```
YScale = 4;
for (int i = 0; i* YScale < YLength; i ++){
    if (i ==0 ||i ==10 ||i ==20 ||i ==30 ||i ==40 ||i ==50
    ||i ==60 ||i ==70 ||i ==80 ||i ==90 ||i ==100
    ){
        canvAndroid Studio.drawLine(XPoint, YPoint -i* YScale, XPoint +5, YPoint
            - i* YScale, paint);
        canvAndroid Studio.drawText(YLabel[i], XPoint -50, YPoint - i* YScale,paint);
    }
}
canvAndroid Studio.drawText(getResources().getString(R.string.company) +
    company,XPoint -30, YPoint - YLength -20, paint);
```

```
canvAndroid Studio.drawLine(XPoint, YPoint - YLength, XPoint + XLength, YPoint
    - YLength, paint2);
canvAndroid Studio.drawLine(XPoint, YPoint, XPoint + XLength, YPoint, paint);
```

- 绘制曲线。实现代码如下：

```
if (swi!=8){
    for (int i=0; i* YScale < YLength; i++){
        if (i==0 ||i==10 ||i==20 ||i==30 ||i==40 ||i==50
            ||i==60 ||i==70 ||i==80 ||i==90 ||i==100
            ||i==110 ||i==120 ||i==130 ||i==140){
canvAndroid Studio.drawLine(XPoint, YPoint - i* YScale, XPoint +5, YPoint
    i* YScale, paint);
canvAndroid Studio.drawText(YLabel[i], XPoint -50, YPoint - i* YScale,
    paint);
} else {
    //绘制曲线
        if (swi==0){
        if (i==75){
canvAndroid Studio.drawLine(XPoint, YPoint - i* YScale, XPoint + XLength,
    YPoint - i* YScale, paint);
canvAndroid Studio.drawText(getResources().getString(R.string.dpm_75),
    XPoint + XLength -20, YPoint - i* YScale -15, paint);
    }
} else if (swi==4){
if (i==57){   //判断数值是否相等
canvAndroid Studio.drawLine(XPoint, YPoint - i* YScale, XPoint + XLength,
    YPoint - i* YScale, paint);
canvAndroid Studio.drawText(getResources().getString(R.string.dpm_37),
    XPoint + XLength -20, YPoint - i* YScale -15, paint);
}
}
}
}
```

```
    } else {
    YScale = 4;
for (int i = 0; i* YScale < YLength; i ++) {
  if (i ==0 ||i ==10 ||i ==20 ||i ==30 ||i ==40 ||i ==50
  ||i ==60 ||i ==70 ||i ==80 ||i ==90 ||i ==100
  ) {
  canvAndroid Studio.drawLine(XPoint, YPoint -i*YScale, XPoint +5, YPoint
      - i* YScale, paint);                              //画文本框
  canvAndroid Studio.drawText(YLabel[i], XPoint -50, YPoint -i* YScale,
      paint);                                           //画文本框
    }
    }
}
canvAndroid Studio.drawText(getResources().getString(R.string.company) +
company,                                                //画文本框
XPoint -30, YPoint -YLength -20, paint);                //计算中心点
canvAndroid Studio.drawLine(XPoint, YPoint -YLength, XPoint +XLength, YPoint
      - YLength, paint2);
canvAndroid Studio.drawLine(XPoint, YPoint, XPoint +XLength, YPoint, paint);
```

● 因为曲线需要动态显示,所以需要不断地更新数据调用 invalidate( )来重写 OnDraw( )函数。实现代码如下:

```
private static clAndroid Studios HearHandler extends Handler {
    private WeakReference <RoundView >mActivityReference;
    HearHandler(RoundView activity) {                   //更新数据
        mActivityReference = new WeakReference < > (activity);
    }
    @Override
    public void handleMessage(Message msg) {
        final RoundView activity = mActivityReference.get();
        if (activity ! = null) {
            if (msg.what == 1) {
                if (activity.data.size() > = activity.MaxDatAndroid Studioize) {
                    activity.data.remove(0);            //删除数据
```

```
            }
                    activity.invalidate();//存储数据日期
        }
    }
}
```

- 在布局 fragment_uv.xml 中声明自定义的 View。实现代码如下：

```
<activity.lsen.wearabledevice.view.RoundView
    android:id = "@ + id/uv_roundview"
    android:layout_width = "match_parent"
    android:layout_height = "0dp"
    android:layout_gravity = "center"
    android:layout_marginTop = "10dp"
    android:layout_weight = "2" />
```

运行后的效果图 5 – 52 所示。

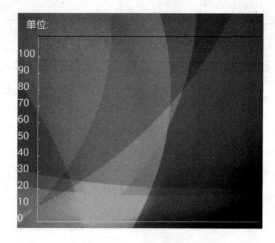

图 5 – 52 运行效果图

- UVFragment 中使用 RoundProgressBar 和 RoundView 在 onCreateView 中使用视图。实现代码如下：

```
public View onCreateView(LayoutInflater inflater, @Nullable ViewGroup container,
    @Nullable Bundle savedInstanceState){
  View view = inflater.inflate(R.layout.fragment_uv, container, false);
```

```
uvRoundview.swi = 8;
uvRoundview.data = pullData;     // 使用 RoundProgressBar
uvNumberEt.setText(getResources().getString(R.string.uv_number));
alarmNum = Double.valueOf(uvNumberEt.getText().toString());
```

- 把收到的紫外线数据设置到视图中。实现代码如下：

```
@Override
public void handleMessage(Message msg){
    final UVFragment activity = mActivityReference.get();
    if (activity != null){
        switch (msg.what){
            cAndroid Studioe 1:                    //紫外线数据设置到视图中
                int value = (Integer) msg.obj;
                if (activity.oldHear != value){
                    if (value >= activity.alarmNum){
                        activity.uvStateTv.setTextColor(0xffffff00);
                        // 把收到的紫外线数据设置到视图中
                        activ.ityuvStateTv.setText(activity.getResources().getString(R.
                            string.uv_state2));
} else {
activity.uvStateTv.setTextColor(0xffffffff);
//设置需要显示的文本
activity.uvStateTv.setText(activity.getResources().getString(R.string.uv_
state1));
}
activity.pullData.add(value);
if (activity.pullData.size() > activity.uvRoundview.MaxDatAndroid Studioize){
    activity.pullData.remove(0);        //删除数据
}
activity.uvRoundview.data = activity.pullData;
activity.uvRoundbar.setCompany(activity.getResources().
    getString(R.string.uv));
double num;
num = value;
activity.uvRoundbar.setProgress(num);    //设置进度条大小
```

```
activity.uvRoundview.swi = 8;
activity.uvRoundview.company = activity.getResources().getString(R.string.uv);
activity.uvRoundview.handler.sendEmptyMessage(1);
activity.oldHear = value;
}
break;
cAndroid Studioe 2:
activity.mainActivity.showToAndroid Studiot(activity.
        getResources().getString(R.string.uv_alarm));
activity.uvNumberEt.setText("");
break;
```

● 根据输入的报警参数,设置紫外线报警。当输入结束后,判断报警参数是否在紫外线范围内,如在则输入成功。实现代码如下:

```
uvNumberEt.setText(getResources().getString(R.string.uv_number));
alarmNum = Double.valueOf(uvNumberEt.getText().toString());
uvNumberEt.addTextChangedListener(new TextWatcher(){
        @Override           //文本字体改变函数之后
        public void beforeTextChanged(CharSequence s, int
                start, int count, int after){
        }
        @Override           //文本字体改变函数
        public void onTextChanged(CharSequence s, int start, int
        before, int count){
        double num = Double.valueOf(CharSequence );
        }
        @Override
    public void afterTextChanged(Editable s){
        String numStr = uvNumberEt.getText().toString();
        if (!numStr.isEmpty()){
            double num = Double.valueOf(numStr);
            if (num <=0 ||num >=100){
                handler.sendEmptyMessage(2);
            } else {
```

```
                    alarmNum = num;
                }
            }
        }
    }
}
```

- 紫外线参数报警判断。当紫外线值大于或等于报警参数时,则提示报警。实现代码如下:

```
if (value > = activity.alarmNum){
    activity.uvStateTv.setTextColor(0xffffff00);
    activity.uvStateTv.setText (activity.getResources ().getString (R. string. uv_
    state2));
} else {
    activity.uvStateTv.setTextColor(0xffffffff);      //设置提示颜色
    activity.uvStateTv.setText(activity.getResources().getString
    (R.string.uv_state1));
}   //提示报警
```

小　　结

本章可穿戴设备模块的设计和开发,以及相关 APP 界面和功能的开发。在学习过程中,可锻炼学生的动手能力,熟练地通过可穿戴设备平台开发相关功能。

习　　题

1. 在体温检测过程中,设备测试的结果,和真实数据是否一样,为什么?
2. 在 APP 开发过程中你遇到的最大难点是什么,怎么解决的?
3. 微型马达和普通马达有什么区别,是否都可以满足本实验的要求?
4. 在紫外线测量中,是否可以用人工设备代替太阳发射的紫外线?